Praise for *Growing Food in a Hotter, Drier Land*

"Gary Paul Nabhan offers a necessary guide to the ways of plants, and to managing water wisely in an increasingly unpredictable climate. Past civilizations could have used a book like this. And if we ourselves don't want to become a distant memory, we would do well to heed the hard-won lessons of desert farmers from around the world, and learn the practical earth skills needed to create a permaculture oasis of our own."

—**Michael Phillips**, author of *The Holistic Orchard* and *The Apple Grower*

"Drylands are home to 40 percent of the world's people: a figure sure to rise in the coming decades as our world grows more parched. That is why Gary Nabhan's latest book is indispensable. Everyone who grows food—make that, everyone who *eats* food—should be grateful he wrote it. An homage to old wisdom and to the latter-day soil magicians who are Nabhan's living muses, it is a rich herbarium of delicious, hardy sustenance and a manual for our future."

—**Alan Weisman**, author of *The World Without Us* and *Countdown*

"In a world where climate change is the new normal, Gary Nabhan offers a blueprint for food production. Using desert agriculture as a backdrop, Nabhan is the ideal guide for understanding and addressing the challenges of rising temperatures, depleting water resources, and ever-shifting conditions. It is a cautionary book of hope, full of dry-farming wisdom, to-do lists, and Gary Nabhan's enjoyable combination of insight and humor."

—**Dan Imhoff**, author of *Food Fight, CAFO,* and *Farming with the Wild*

"In *Growing Food in a Hotter, Drier Land*, Gary Paul Nabhan has crafted a cogent treatise blending his own considerable knowledge and experience with the traditional ecological wisdom of indigenous desert farmers, who have been thriving in the face of climate uncertainty for many generations.

"The hard-won lessons and innovations described in this book are applicable for farmers cultivating in all changing climates, and inspirational for all people who depend on their survival and success. A must-have arrow in the quiver for all pragmatic Thrivalists!"

—**Brock Dolman**, director, WATER Institute and
Permaculture Design Program, Occidental Arts & Ecology Center

"We face an unprecedented future. The scale and speed of the changes bearing down on us as a consequence of climate uncertainty has no analog in history. Fortunately, we have guides like Gary Paul Nabhan to lead us through the crazy labyrinth in which we find ourselves. By looking to age-old practices and taking lessons from nature, Dr. Nabhan builds a compelling case for a type of resilience that matters whether you are a food producer or eater—which is everyone!"

—**Courtney White**, founder and creative director, Quivira Coalition

"All of Gary Nabhan's books carry us on deep, enchanting journeys to the hearts of people, plants, and cultures across the world. *Growing Food in a Hotter, Drier Land* offers the rich stories and cultural insights we've come to expect, but now, when we badly need it, Gary also tells us explicitly how to use the dryland wisdom he's assembled over a lifetime. Heaped with practical principles, techniques, plant lists, parables, and more, his new book offers important tools for preserving our food and water security on a warmer, stormier planet. I'm inspired and heartened by this timely and important offering from a true desert sage."

—**Toby Hemenway**, author of
Gaia's Garden: A Guide to Home-Scale Permaculture

"If the 20th century strove to insulate us from the harsh realities of nature (while exacerbating its extremes), Gary Nabhan's latest book introduces us to the 21st century's rude reminders that change is here, uncertainty commonplace. With little room for the hand-wringers, Nabhan provides everyone else, from novice gardener to deep ecologist, important food for thought and the practical know-how to address our modern problems with ancient desert wisdom. I couldn't put it down."

—**Richard McCarthy**, executive director, Slow Food USA

"Gary Nabhan's books never fail to inspire and inform me. This book is no exception. After just one read through, I've dog-eared, highlighted, and noted countless gems, facts, and stories to which I will return again and again. The pattern of the book makes this easy. Each section begins with a Warm-Up problem, followed by a Parable of people or natural systems addressing the problem. Principles and Premises distilled from the problem and parable, along with Planning and Practice tips, then help me work cooperatively *with* the life around me to formulate solutions unique to my site's conditions and changing climate.

"Best of all, I feel I'm part of an incredibly diverse, caring community as I do so, thanks to Gary sharing so many engaging examples of different people, cultures, and ecosystems doing likewise. Read this book!"

—**Brad Lancaster**, author of *Rainwater Harvesting for Drylands and Beyond*

Growing
Food in a
Hotter,
Drier Land

Growing Food in a Hotter, Drier Land

Lessons from Desert Farmers on Adapting to Climate Uncertainty

GARY PAUL NABHAN

Foreword by Bill McKibben

Chelsea Green Publishing
White River Junction, Vermont

Project Manager: Hillary Gregory
Developmental Editor: Benjamin Watson
Copy Editor: Laura Jorstad
Proofreader: Helen Walden
Indexer: Peggy Holloway
Designer: Melissa Jacobson

Printed in the United States of America.
First printing May 2013.
10 9 8 7 6 5 4 3 2   14 15 16

Our Commitment to Green Publishing
Chelsea Green sees publishing as a tool for cultural change and ecological stewardship. We strive to align our book manufacturing practices with our editorial mission and to reduce the impact of our business enterprise in the environment. We print our books and catalogs on chlorine-free recycled paper, using vegetable-based inks whenever possible. This book may cost slightly more because it was printed on paper that contains recycled fiber, and we hope you'll agree that it's worth it. Chelsea Green is a member of the Green Press Initiative (www.greenpressinitiative.org), a nonprofit coalition of publishers, manufacturers, and authors working to protect the world's endangered forests and conserve natural resources. *Growing Food in a Hotter, Drier Land* was printed on FSC®-certified paper supplied by CJK that contains at least 10% postconsumer recycled fiber.

Library of Congress Cataloging-in-Publication Data
Nabhan, Gary Paul.
  Growing food in a hotter, drier land : lessons from desert farmers on adapting to climate uncertainty / Gary Paul Nabhan ; foreword by Bill McKibben.
    p. cm.
  Includes bibliographical references and index.
  ISBN 978-1-60358-453-1 (pbk.) — ISBN 978-1-60358-454-8 (ebook)
  1. Arid regions agriculture. 2. Crops and climate. 3. Climatic changes. I. Title.

  S613.N33 2013
  631—dc23
                    2013008070

Chelsea Green Publishing
85 North Main Street, Suite 120
White River Junction, VT 05001
(802) 295-6300
www.chelseagreen.com

MIX
Paper from
responsible sources
FSC® C103722

Contents

If there was ever a moment for this book, now is it. In 2011 we saw record heat across Texas and Oklahoma—the hottest summers ever recorded in the US, helped along by a drought so deep it killed 500 million trees. In 2012 that heat and drought spread across the heart of the nation, driving crop yields down to lows not seen for decades and setting a new high-temperature mark for the entire United States. If you've reached the point where you can't grow corn in Iowa—in the richest soil on earth—then you've reached the point of real trouble.

And of course more to come: we've raised the planet's temperature a degree so far, but that's just the start. Unless we get off coal and gas and oil far quicker than any government currently plans, the temperature will rise 4 or 5 degrees this century, which is to say past the point where agronomists think we can support the kind of civilizations we now enjoy. Even in the best-case scenarios, though, agriculture is going to get much harder than it is at present—that's what those scenes from the Midwest last summer demonstrated.

To cope, we'll need more hands on the farm. And Gary Nabhan, with both beauty and precision, demonstrates exactly why. He has a dozen wise prescriptions in this book for how we might be able to keep growing food even in very harsh places. But all of them demand people out there working with their hands: building the "fredges," sinking the clay jars into the soil. We'll need to interplant and intercrop and shade—and none of that can be done by one farmer piloting a giant combine across a ten-thousand-acre sea. It will have to be done by caring human hands, connected to very smart and nimble human minds.

The good news is that those humans are starting to appear. In the last few years, for the first time in a century and a half, the US Department of Agriculture has reported that the number of farms in the US is increasing instead of shrinking. There are plenty of kids each year now who head out of WWOOF (Worldwide Opportunities on Organic Farms) to work on organic farms around the planet, weeding and planting in return for room, board, and a serious education. Local food networks have begun to spread around the nation: the farmers' market has been the fastest-growing part of the food economy for a decade now.

Whether it will happen in time to catch up with the physics of climate change is an open question. I watched in 2011 as Hurricane Irene dropped record amounts of water on Vermont. Among the casualties were plenty of our most hopeful small farms, their rich fields turned into rocky stubble. Whether you look at the hydrological projections for the Colorado River or for the Jordan River, it's hard to see how it's all going to work out.

But the basic human question, for as far back as we can imagine, has always been: What's for dinner? We'll try, as best we can, to make sure the answer is: something. And as this entirely lovely volume makes clear, there will be beauty as well as despair in the solutions and fixes. We'll be forced away from relying on the almost military might agribusiness has brought to bear, the attempt to overwhelm nature with chemicals and fossil fuel. Instead, as Nabhan says, we'll need to understand nature with more precision and insight, mimicking the things that it does with such unrelenting vigor.

We've thought ourselves wise for several generations now, but in fact that wisdom has been a simplifying kind. Now we're going to need exactly the kind of complex, place-based wisdom that Nabhan outlines here. We're going to have to wise up, in a hurry. And the biggest part of that wisdom will involve realizing that we depend on others. On other farmers who can help us, and on other species whose aid we'll require, too. Some mixture of humility and new pride just might see us through—that's the note to take from this book, I think. That and a thousand good ideas about what to do as the growing season begins!

Bill McKibben
Middlebury, Vermont
February 2013

Acknowledgments

I first and foremost want to thank Chelsea Green's Ben Watson and Joni Praded for inviting me back to do another book with their fine team. I am also indebted to the W.K. Kellogg Foundation and the Agnese Haury Fund for supporting my work through the University of Arizona's Sustainable Food Systems program at the Southwest Center. I owe deep thanks to my Franciscan spiritual mentors—from Nancy Menning to Richard Rohr—who encouraged me to tell a fresh story of human responses to climate change, rather than dwelling in the same thought patterns that created and aggravated the problem in the first place.

After my initial inspiration to be engaged in climate change adaptation issues through the fine writing and friendship of Bill McKibben, William deBuys, and Julio Betancourt, the university's Institute for the Environment and Udall Center for Public Policy provided me with a fellowship to focus on how climate adaptation could enhance our food security. The Tumamoc Hill's Lab for Plants and People and the National Phenological Network, both based in Tucson, have also offered extraordinary support and learning opportunities with respect to long-term phenological changes triggered by global change and the urban heat island effect. Tumamoc's artist-in-residence, Paul Mirocha, assisted by Barbara Terkanian, has, as usual, come through with extraordinary illustrations that honor the desert and its diverse people. David Cavagnaro, Josh Dick, Laurie Monti, Chet Phillips, and Caleb Weaver kindly offered photos to supplement my own. In addition, Caleb Weaver, Nate O'Meara, Kanin Routson, and Regina Fitzsimmons either assisted me with field demonstrations, helped with data compilations, or read earlier drafts of this manuscript; their youthful energy, curiosity, and skills greatly enhanced the depth and breadth of this work.

What a remarkable chance this has been to recap much of what I have learned as a desert naturalist and part-time orchard-keeper from my interactions with traditional desert farmers around the world. In particular, I am honored to have known Jerome Ascencio, Ogonazar Aknazarov, Aziz Bousfiha, Jim Corbett, Adalberto Cruz, Howard Scott Gentry, Fred Kabotie, José Kerman, Laura Kerman, Delores Lewis, Humberto Romero Molina, Edna Tallas, Leon Tsosie, Vernon Masayesva, and many other desert farmers or ranchers. I have also learned from many peers who are just as extraordinary: Miguel Altieri, Juan Estevan Arrellano, David Bainbridge, Mary Bender, Julio Betancourt, David Cavagnaro, David Cleveland, Bill Doolittle, Richard Felger, Nancy Ferguson, Suzanne Fish, Wes Jackson, Greg Jones, Fred Kirschenmann, Brad Lancaster, Anna Lappé, Nancy Laney, Diana Liverman, Eike Luedeling, Bill McDorman, Bill McKibben, Suzanne Nelson, Tom Orum, Michael Phillips, Humberto Romero Morales, Rafael Routson, Tracey Ryder, Chris Schmidt, Tom Sheridan, Evan Sofro, Daniela Soleri, Bill Steen, Nancy Turner, Peter Warshall, Jesse Watson, and Ken Wilson.

Finally, my wife and arid lands ethnobotanist Laurie Smith Monti has been gracious enough to share the road with me across many deserts and through many droughts on our own lands. I am graced by her love and her passion for life every day. Who else but Laurie would be willing to go with me to the Gobi and the Sahara in the heat of the summer and ride mules with me into canyon oases in Baja California in the cold of the winter? That's just an inkling of why I am so grateful for my *compañera* . . .

– Introduction –

Wasteland or Food-Producing Oasis?

A Time to Choose

Warm-Up

As I begin this effort to share with you the many remarkable ways that desert farmers around the world have adapted to their changing climatic conditions—conditions that global climate change will be bringing to many more **foodscapes**[*] on this planet—I want to take you back to where the journey began for me. Perhaps you will then understand my motivations for gathering these stories of adaptation and resilience, and why they might inspire all of us to improvise similar strategies on our own home ground.

For much of the last four decades, I have lived in the Sonoran Desert of southwestern North America, working on both sides of the US–Mexican border. The farming cultures living in this arid region—like traditional communities of farmers in deserts elsewhere—have experienced at least 700 years of adaptation to dramatic climate changes while continuing to farm in many of the same places their ancestors did.[1] I have had the privilege of visiting and working with traditional farmers in other arid and semi-arid regions as well: the Sahara Desert; the Omani highlands above the Empty Quarter of the Arabian Desert; the Gobi, Taklamakan, and Pamirs deserts of Central Asia; the Painted Desert of the Colorado Plateau; and the Chihuahuan Desert. While most of the farmers I have spent time with have lived their entire lives within one particular desert, I've realized that few, if any, of them viewed the arid conditions they face

---

[*] Terms that appear in bold are defined in the glossary at the back of this book.

today to be identical to those they encountered when they were children growing up in the very same landscape.

As you shall see or hear in the following chapters, oral histories kept by indigenous farmers from around the world suggest that conditions today may be hotter, drier, or more uncertain than were found in previous generations. (In chapter 3, for instance, you will hear the peculiar story of the first Sonoran Desert farmer who told me that "the rain is dying" where he lived, well before scientists convinced me that long-term rainfall patterns in our region had been shifting with global climate change.)

When I was younger, I was at first skeptical of such accounts, assuming that land degradation—not changes in rainfall—was the reason that my friends had found it increasingly difficult to farm as compared with the past. And yet for another three decades I continued to record accounts from traditional desert farmers that suggested dramatic shifts in weather over their lifetimes, shifts that disrupted the way their communities had farmed for decades, if not centuries.

After years of farmers hinting to me that climate change was affecting their ability to farm, I rather suddenly moved up to the Grand Canyon region to try my hand at **dry-farming** and sheep raising on the flanks of the San Francisco Peaks. There I initially assumed that rain- and snowfall would provide adequate moisture for the small field and sheep pastures my wife and I had purchased, compared with what we had experienced 30 miles outside Tucson in the Sonoran Desert.

As it happened, I couldn't have been more wrong.

Ironically, we had arrived on the Colorado Plateau just after the onset of one of the worst droughts and insect outbreaks in the region's recorded history. Within a matter of eight years, I witnessed more rapid change in a single landscape than I had seen at any previous time during my life. Roughly 80 percent of the ponderosa and piñon pine trees within the square mile around our home died of bark beetle infestations. The population buildup of beetles had been aggravated by drought and mild winters, which allowed multiple generations of the pests to move from tree to tree.

Every time I sowed and watered native pasture grasses on our now-treeless land in order to provide more forage for our sheep, the seeds would germinate, but their seedlings would emerge then quickly die from the blistering heat that accompanied the drought.

In the end, we had to resort to purchasing hay virtually year-round to keep our flock of hardy Navajo-Churro sheep alive during the drought. Hay prices in the region had more than doubled since 2000, due to the scarcity of water for use in irrigating crops there. In the hay-producing valley closest to our home, 80 percent of the springs and seeps had gone dry. We did not know at the time whether the regional drought could be directly related to long-term climate change, but it seemed like a harbinger of what was to come. Rather than further degrade a thirsty land with a flock of sheep it could no longer support, we decided to move back down to the edges of the Sonoran Desert, where I at least knew how to select crop varieties that were well adapted to conditions there. We left the Churro ewes and lambs with Leon Tsosie, a Navajo friend and sheep-herding mentor who had far more pasturage and water (not to mention traditional knowledge!) around his village with which to support a flock.

And yet the first summer we landed back in the Sonoita Creek drainage where I had lived in southern Arizona some 35 years earlier, I witnessed something I had never observed at all over the intervening decades. One light rain briefly dampened the dust in early June, but at the time when the summer monsoons would typically begin a few weeks later, no rains came at all. In fact, the grasslands of the Sonoita Plains *never* greened up that summer. My neighbors—mostly ranchers—sold off half or more of their herds, and pulled the remaining livestock off the most brittle range to keep in well-watered pastures. The plants in our new little garden were repeatedly damaged by wildlife, which had little to eat out on the surrounding range.

It was then that I conceded that I could not escape climate change, no matter where I moved. I resolved myself to accepting that it would bear down on us for the rest of our lives, and I began to grow despondent and uncertain as to whether I even had the capacity to produce food anymore. I also came to reluctantly acknowledge that, despite the hard work and intelligence most farmers and ranchers bring to producing food for Americans, our food system as a whole—from the purchase of energy-intensive inputs, to production, harvesting, processing, distribution, and preparation—has played a major role in generating the greenhouse emissions over the last century that have accelerated climate change.

As I read more and more revelations about the accelerated rate of climate change, written by some of the most trusted scientists I had known over my career as a desert agro-ecologist, I began to despair, convinced there was little that we could do over our lifetimes to stop the planetary hemorrhaging. My friend Bill McKibben has echoed their concerns, reminding us that while greenhouse gases have already raised the temperature of the earth by 2°F (1°C), it is predicted to rise another 7 to 9°F (4–5°C) by the end of this century if nothing substantive is done. McKibben has noted that current models demonstrate how each (Celsius) degree of global warming could potentially cut grain yields by 10 percent, which would have devastating effects on food security for all of humankind. And my friends in conservation seemed to be writing one obituary after another for river after river, as well as penning terminal diagnoses for populations, species, communities, cultures, and landscapes now endangered by severity of climate change. At the same time, my friends in the arts—from Alison Deming and Gretel Ehrlich to Terry Tempest Williams—were also spending much of their time bearing witness to the causes and consequences of climate uncertainty. I had not set out to become a desert naturalist or an agricultural ecologist merely to exhaust myself and others with fatalistic diagnoses, obituaries, or autopsies of what I loved about the world we live in. Yet for a season or two, I felt incapacitated from imagining, let alone taking, any positive action in response to climatic disruptions, especially when it came to how shifting weather patterns would affect our food security. I sensed that, to get out of my depression, it would take me hearing a fresh story—one that might give me an altogether different way of thinking about and acting in response to climate change.

Parable

Fortunately, around that time I met a remarkable Moroccan farmer and Sufi visionary, Aziz Bousfiha,[2] who lived on the arid edge of Fez, the World Heritage Site historically inhabited by Berber, Arab, and Sephardic Jewish artisans, scholars, scientists, and holy men. Some mutual friends of ours thought that I would be interested in visiting the remarkable orchard-gardens that Aziz had planted around his parents' home, and to hear of his efforts—already in process—toward restoring an ancient oasis out beyond the last suburban settlements of the sprawling metropolitan area. Fez itself is a semi-arid landscape, historically averaging just less than 20 inches of rainfall a year, but the aridity of the Moroccan landscape increases dramatically as you travel south of the city, toward the western edges of the Sahara.

Indeed, as I came out of the glaring sunlight of a Moroccan thoroughfare into the shade and fragrances of the Bousfiha family estate, it was clear that someone there—probably Aziz himself—had the gift of a green thumb. There were trees filled with jujubes and loquats, pomegranates, and prickly pear cactus fruits. There were kumquats, grapefruits, mulberries, agaves, and olives. As I wandered through the orchard before going into the house, I also caught the scents of lavender and rose, limes, and sweet lemons. But that was not all. Below these fruit trees there were basins filled with scallions, onions, leeks, and garlic. Between the fruit trees, there were rows of the crocus flower from which saffron is extracted, as well as coriander, sage, spearmint, peppermint, rosemary, anise, oregano, thyme, and epazote. Although it was late in April, Aziz still had plenty of beets, radishes, cabbages, and fava beans left from his winter plantings. But he had also recently put in his plantings for the summer, including sunflowers and safflower, chile peppers and tomatoes, as well as corn, beans, and squash.

This, I thought, *is an urban oasis*—a lush patch of green in the midst of a Moroccan metropolis that was already suffering from the **urban heat island effect** of higher nighttime temperatures and patchier rainfall due to the high density of buildings, asphalt-covered roads, heaters, and even air conditioners.

But what I did not know then was that the term *oasis* meant far more to Aziz Bousfiha than anything I could describe about the diverse greenery found on his family's estate. After taking off our shoes, and greeting his father, a learned and esteemed member of the Islamic community of Fez, we were escorted into the house to sit and drink tea with Aziz himself, who was on crutches as the result of a recent fall. Despite the pain he was still suffering, Aziz welcomed us in halting but beautiful, soft-spoken English. He then tried to offer us his vision of what an oasis—or a chain of them around the world—could ultimately mean to humanity.

Listening, I leaned back against a cushion, floored by what Aziz had not only imagined but begun to build in the face of climatic uncertainty:

"For me, the idea is to go somewhere into the desert [to find a place to share with others—one severely degraded over time by neglect, depletion of water, and perhaps climate change]. We'll proclaim that yes, this place has been **desertified**, but now we're going make it into a *living oasis*, one where we will respect and nurture a diversity of life."

"In fact—" He smiled, drank some tea, then began again. "—we'll arrest [or bring to a halt] all activities and uses of chemicals that deplete diversity. If pesticides have been shown to kill any species, we'll eliminate their use.

"Next, we will return to the adapted seeds of the region [to restore diversity in the oasis], for they play multiple roles, whereas a modern hybrid seed plays only one or a few roles. When we narrowly manipulate seeds to give us more yields in quantity [but not quality], I see only one sad goal: driving the [food production] system toward profit . . . To counter this trend, we must remember that centuries of love and care are

Sufi mystic Aziz Bousfiha at work toward his vision of a resilient oasis culture here on earth, restoring an abandoned ancient oasis near Fez, Morocco. PHOTO BY JOSH DICK.

manifested in each kind of little seed. There is love and care embedded in each olive growing on our trees, in each grain ground into our bread.

"But we don't stop with making [this diversity flourish at] just one oasis. The idea is to bring many small oasis-like farms into a chain [a corridor, a connected network or community], so that all of them will be in exchange with one another in order to serve the larger community of which they are part.

"They will also be in exchange with the more urban settlements nearby that cannot grow all of their own food. And so the farmers [from the desert oasis] will give a portion of the food they produce to urban dwellers, who will become *ambassadors* for them.

"We will generate solidarity between people on and off the farm, who will begin to walk the long road of ancient wisdom together. They will bring back the old grains of the region as symbols of the seeds of wisdom that we must plant. Over the centuries, these ancient seeds have [dynamically] adapted to *place*. It is not just a natural ecosystem, but a cultural ecosystem as well.

"It is not just about farming with plants [adapted] to a particular desert oasis, it is about cultivating solidarity among people . . . as well as *a thousand other things that will emerge from that.*

"And so we must regenerate a series of diverse desert oases, and then link them through our solidarity. That solidarity—not just [what happens in] any one place—is the cohesiveness, the core which will keep all this diversity intact. Each place, each oasis, is [in and of itself] important. But no single place is more important than the others. We have to be more than just brothers [and sisters], each doing our own thing. We must work and pray as members of the same community."

There was silence. Those of us assembled together with Aziz were speechless, humbled, and stunned. Finally, one of us asked if he thought such a vision for creating a more resilient and desert-adapted agriculture could fly in the face of climate change. I sensed I knew what his answer might be, and I was close.

Aziz laughed gently, and then replied, "I can't waste time worrying about whether or not this will work. There is a proverb in Arabic—and [probably] similar ones in other languages—that may say it all: If it looks like the last day of the world is upon us and the end of life may be coming . . . and you realize this moment while you are planting trees, well, *don't stop planting!*"

We laughed along with him, and then another silence took hold of the room. After a minute or two, Aziz stood up high above those of us still sitting on the floor. He began gesturing, but as his arms and hands moved around, I absentmindedly mistook them for limbs, for the branches of a tree blowing in the breeze, keeping itself balanced and rooted in the face of a terrible storm.

"It's not just activism I am talking about, although we may [indeed] need that. I am talking about something larger, deeper: *participating in the creation*—for that is the [ultimate] expression of our love. Yes, love is the driving force, I can say that. *Love, why not?*

"Every day, I go out to prune and renew [the productivity of] 1,000-year-old olive trees on the land that has been gifted to me [at that degraded old oasis]. I must take care of them with love, not just with the science of pruning . . .

"So please join me," Aziz concluded, gesturing for all of us to get back on our feet. "Let us go out into this little oasis, and I'll show you some of what may come out of all of this . . ."

Principles and Premises

Since that moment with Aziz, I have never been able to think about our fate in the face of climate change in the same way. I realized that this is a time ripe for creative action, not just for scientific analysis.

We will need more than science to help us prune back our gangly, overgrown carbon footprints, which are driving the anthropogenic portion of the climate change equation. We will need to care for the diversity of this world as if we truly love it, and not merely because of the utilitarian fact that such diversity tangibly nourishes us. We will need to take back the farms, orchards, ranches, fishponds, and gardens most degraded by climate change—ravaged by drought, heat waves, and blight—and patiently find ways to heal them so that they are sanctuaries where springs cannot be overtapped and food diversity is enhanced, not diminished.

This book builds on the vision of farmers like Aziz Bousfiha who want to ensure a resilient food future is made manifest in the new world that is emerging—a world of profound uncertainty that scientists refer to as "non-stationarity."[3] Aziz has offered a vision that can and should empower us, instead of letting fatalism, cynicism, and environmental determinism further impoverish us. But it must empower us toward collective action—to redesign our entire food system, to reduce its carbon footprint and increase its resilience—rather than thinking that it is enough to merely make a single garden, farm, or ranch more sustainable.

What Aziz portends—and what each of us can affirm that we are willing to do—is to become co-designers of an alternative food network, one that:

- Slows and mitigates climate change;
- Reduces carbon consumption, post-harvest waste, and the size of our ecological "foodprints";
- Builds more structural resilience through plant, animal, and microbial diversity to buffer us from catastrophic events;
- Relocalizes food production, processing, distribution, and consumption on a culturally responsive scale; and
- Empowers people from all cultures and livelihoods to actively and passionately participate in community-based process and policy changes that build toward true food democracy and climate justice.

What might these alternative food networks look like and taste like? Well, because true solutions are place-based *and* community-based, there is no cookie-cutter approach that can possibly work in all places for all peoples, given the complexities of our food systems—even if we do not fully take into account the multiple consequences of climate change.

Just why won't cookie cutters work? Climate adaptation is inherently *place-based* adaptation, however much it must be set in a global context of cause and effect. To work, you and your neighbors must be fully and creatively part of the adaptation process. Top-down "solutions" won't fly; they are more likely to crash. However much I love James Workman's *Heart of Dryness* book about "how the last Bushmen can help us endure the coming age of permanent drought," his tongue-and-cheek question, "What would Bushmen do?" doesn't go far enough.[4] "What would you and your neighbors do?" is the more critical question. Bottom-up means that you and your neighbors must stay *more attentive to where you live and grow food*, and become more engaged with other members of your ecological and cultural community. We must not attempt to imitate, but instead emulate what traditional desert peoples have learned about living with climatic uncertainty. Take inspiration, not recipes.

Let me add one additional caution: While I use the *wisdom of the desert* as a metaphor and means to stimulate innovation at many points in this book, climate change will not necessarily make most places on the face of this earth look like true deserts. Desert ecosystems have their own integrity, just as prairies and rain forests do. While hotter, drier conditions will come to be commonplace in many (though not all) temperate and tropical landscapes, there are also other changes that will occur: carbon enrichment of the atmosphere, depletion of fossil fuel and water reserves, and so forth. The integrity of prairies, deserts, wetlands, and forests may become compromised in unseen ways. The important point here is not that all desert adaptations will work in other landscapes, but that they can serve as inspirations for local adaptations that do.

Planning and Practice:
Know Yourself Through Knowing Where Your Relationship with Your Foodshed Is Flowing

In each chapter of this book, I will initiate a set of pragmatic responses to each parable offered by a desert elder. The first set of responses will focus on what food producers should be planning for, and the second will be engaged with on-the-ground responses that we can all activate in some way in our fields, orchards, ponds, gardens, and kitchens.

In essence, all of us must learn to plan for uncertainty in the food system. While that expression, *plan for uncertainty*, may seem paradoxical, we should remember the advice that Canadian Walter Gretsky once offered his son Wayne, who went on to become the highest-scoring hockey player in history, "Go to where the puck is going, not where it has been."

As my friend Bill DeBuys, author of *A Great Aridness*,[5] has reminded me, what works as a guiding principle for ice hockey players might serve just as well in an arena with little or no ice, that is, the arena of food production under increasingly desert-like conditions. Here then are the questions you must ask yourself as a food producer and as eater as we begin to imagine ways to redesign our foodscapes.

- What conditions will you face as you produce food in the future—conditions that may be altogether unlike those you have dealt with in the past?
- What five words describe the values that you want to see embedded in your foodscape, and manifested in your harvest?
- If the late Wallace Stegner was correct that we must "get over [our addiction to] the color green," what other colors do you wish to add to your foodscape?
- If you are successful in adapting your food production, distribution, and processing practices to make them more climate-friendly and resilient, what will your successes taste like?
- What old behaviors, technologies, and practices might you have to give up or suspend to get to where you want to be?
- How will your life—and the lives embedded in your food-producing land—be richer and healthier if you choose to make these changes?
- What kind of social, technical, and financial support will you and your community require to bring such a vision to fruition?

I am writing this book not merely with the hope that you will begin to think about climate change and food in a fresh way, but that you will *act* in new ways as well. This book is as much about *praxis*—taking right action—as it is about *gnosis*—acquiring appropriate knowledge. In each chapter, I will ask you to look at the land and water from which you acquire your food, and to take steps to shift your relationship to it.

Over the years, I have been deeply influenced by Jim Corbett, the late desert visionary, goat-walker, gardener, rancher, and human rights advocate who helped to found the Sanctuary Movement in the early 1980s.[6] A few years before he formed an underground railroad to help political refugees from Central America take refuge in the United States, Corbett and I were in the same gardening and goat-milking co-op in Tucson, Arizona. Near the end of his life, Corbett and his neighbors living in the rural community of Cascabel collectively elaborated and executed an ecological and ethical contract with "their" land that has become known as the Saguaro-Juniper Covenant. Each of us can and should make our own particular pact with the land in which we live; mine is posted as the "Credo" for our foodshed on the website, www.garynabhan.com. To begin to make the shifts needed to survive climate change, I suggest that you gradually put into practice the following behaviors and principles:

- Walk the land you derive your food from, and keep a journal of how it is affected by the way your food is currently grown and eaten; in addition, note any tangible signs that climate change is affecting its capacity to produce food.
- With your friends and neighbors, discuss how the future of your health and the land's health may be affected by climate change, and how your own land management and consumption patterns may be contributing to the drivers or "causes" of climate change.
- Map the places in your foodscape that you expect may be most vulnerable to climate change, and list three ways you might reduce their vulnerability or adapt to those changes.
- Map how water has flowed across or around your land during both wet years and dry years, and compare those two maps, highlighting the differences.
- Note what has been killed, damaged, or degraded by recent weather-related events, and what might have been spared if stewardship practices had been different.
- Make a shrine on or near the site(s) of one of those deaths, degradations, or losses of diversity, as a tangible benchmark of change occurring in your midst.
- Map your neighboring foodscape and your larger **foodshed**—that is, where within sight of your home there are wild or managed habitats where food is produced, not only on your property but on that of neighbors. Then extend your map out to where other significant quantities of your food come from.
- Identify from the foods located on that map where there might be possibilities for food exchange, sharing, or mutual support among your neighbors.
- List five actions that you can gradually begin to enhance the food-producing capacity of your neighborhood or community foodscape.
- Over the following weeks, initiate at least three meetings with your neighbors to collectively write a covenant with the land that engages

all of you in helping your foodscape become an oasis or wellspring for long-term food-producing possibilities, and then take tangible actions toward that shared vision.

Let's get started!

– Chapter One –
Getting a Grip on Climate Change:
Crossing the Threshold into Chronic Climatic Disruption of Food Security

Does the desert's infinite vastness terrify you? Then it would do you well to remember: The desert loves you . . . it loves to strip you bare.
—SAINT JEROME, early desert hermit in the
Nitrian Desert of Egypt AD 385–388

Warm-Up

Weather and food go hand in hand, like beans and corn bread or biscuits and gravy. And so it should come as no surprise that the health and security of our foodscapes is a sensitive indicator of our resilience to the vagaries of climate change. Recent climate-related disasters remind us that our modern, "conventional" means of growing and distributing food have become ever more vulnerable as climate change accelerates. In fact, industrial agriculture and its extra-local distribution systems are increasingly challenged by climatic disasters, as the news media report to us more and more frequently.

In August 2012, the United Nations Food and Agriculture Organisation (FAO) reported that droughts across the Americas had caused global food prices to jump 6 percent in a single month, with the global prices of corn soaring 23 percent that previous July. If this were the only month in recent times that the weather threw farmers a curveball, we could forget about it and go on with the game. But as Arif Husain of the UN Food Programme recently warned, "The real problem is [that] this is the third price shock in the last five years. The poorer countries haven't had time to recover from the previous crises."[1]

Agrarian communities in steep mountain ranges such as the Pamirs in Tajikistan have already been differentially affected by climatic shifts, with changes in the growing season, winter chill hours, frequency of landslides, and disrupted water availability threatening their food security.

Of course, climate change has not descended upon agrarian landscapes and rural communities in one fell swoop. Nevertheless, a certain sort of agricultural history was made in 2011 when more than 500 food-producing counties in the continental United States were declared parts of disaster areas because they suffered weather-related crop failures. The searing heat waves and drought suffered across seven-tenths of the United States during the summer of 2012 have been even more devastating, with 2,228 counties declared part of disaster areas from having crops and livestock damaged by drought well before the devastating floods of Superstorm Sandy hit the East Coast. In 2012, 40,000 new daily records for hot temperatures were set across the country, and 21 million people lived through extreme heat advisories. Furthermore, 2.1 million acres of wild lands, farmlands, and residential communities were burned to a crisp by wildfires. These are further signs that the earth's temperature regulators are not necessarily functioning as they have in the past.

Whatever the current year offers your region in terms of climatic disasters, for me, the threshold into the "new normal" was clearly crossed in 2011, when 99 major disaster areas were declared by President Obama,

A Hot Time in Texas?
NASA Links It to Longer-Term Climate Change

Is a lingering drought necessarily evidence of long-term climate change? No, not in every case, but recent climatic disasters affecting ranching and farming in Texas can indeed be explained by longer-term trends. In 2011, heat spells, drought, and associated wildfires generated a record $7.62 billion in crop and livestock losses in Texas alone. Millions of usually dry-farmed acres never even received enough rainfall to germinate the seeds that farmers planted that year. Ranchers ran out of forage in their pastures, hay in their barns, and money in their bank accounts when the prices for hay imported from other states tripled overnight. Dr. David Anderson, a livestock economist with Texas Agri-Life Extension Service, tried to put in perspective the magnitude of climatic disruption, which reached $3.2 billion of losses in the Texas meat industry well before 2012's countrywide drought hit parts of Texas for a third year in a row:[2]

"2011 was the driest year on record and certainly an infamous year of distinction for the state's farmers and ranchers. The $7.62 billion mark for 2011 is more than $3.5 billion higher than the 2006 drought loss estimates, which previously was the costliest drought on record. The 2011 losses also represent about 43 percent of the average value of agricultural receipts over the last four years."

Analyses done by NASA's James Hansen, director of the Goddard Institute, left no doubt in scientists' minds that the drought affecting Texas was tightly linked to long-term climate change. He suggested that without long-term climate shifts, there would have been an extremely low probability of lasting drought in the South and the Midwest, or in northern Mexico, where 1.7 million head of livestock died on the range. In calling the 2011 drought the worst to hit the US/Mexican desert borderlands in half a century, Mexico's president was pretty close to the call made in NASA's 50-year analysis of land and air temperature records from that region. From NASA scientists' view, human-induced activities have so loaded the earth's climate that the extreme temperatures experienced in North America in 2011 were beyond the range of natural climatic variability "because their likelihood was negligible prior to the recent rapid global warming."[3]

One of 1.2 million cattle killed in northern Mexico by the 2011–2013 drought that ravaged food production in eighteen US and seven Mexican states.

Many perennial streams, as well as the diverse wildlife in their riparian habitats and the irrigated fields and orchards benefiting from them, are now endangered by climate change.

breaking all previous records. Although not every disaster could be attributable to long-term climate change that year—for instance, the East Coast earthquake was not—the severity of climatic disruption to agriculture in 2011 was unprecedented. Furthermore, the National Oceanic and Atmospheric Administration (NOAA) reported that, for the first time in American history, emergency services and cleanup efforts alone topped $35 billion in losses to the country's economy. While the number of disaster designations in any given year may be somewhat affected by politics, the true costs of emergency responses and mop-up are not.

While crossing the 500-county threshold of climatic disruption in 2011 may have served as wake-up call for some of us, many Americans seem oddly oblivious to the magnitude of the shifts in our food-producing capacities that have already occurred. And yet, even when 2,300 counties were declared parts of national disaster areas because of the direct affects of devastating drought, floods, hurricanes, or wildfires in 2012, it seemed that policy makers treated these catastrophes as somehow independent of long-term climate trends, or of the human capacity to affect and be affected by climate itself! At least, by late winter, 2013, as meteorologists conceded that the drought would be persisting for several more months in parts of every state west of the Mississippi River, more attention was still being given in Congress to Superstorm Sandy payouts than to the long-term causes, consequences, and costs of drought and warming trends.

Given the societal responses to date, this book is meant to empower you as readers to be alert to climate change right where you live, and to adapt to it through the way you access food. Just as most of our fellow citizens may be out of touch with long-term weather trends around their home or farm, or along the closest waterfront, many remain relatively

Even in the heart of the Sahara Desert, near Siwa, Egypt, heat-tolerant crops can be grown if water is carefully managed.

Mountain streams that supply irrigation water to farmers in the Shokdara Valley of Tajikistan are seeing the timing and volume of their flows affected by climate change.

unaware of where their food comes from. Twenty-four straight days of temperatures exceeding 100°F (38°C) in Houston or 50 days of the same in Waco may make the front page, but the impacts of longer-term warming trends in the Southwest on farms, ranches, and fisheries hardly ever stay in the headlines or coffee shop conversations for very long.

Of course, it is not just the vast ranches and dry-farmed grain fields of Texas and northern Mexico that now have a heightened vulnerability to climatic disruption. All around the world, many kinds of food production are already suffering from the effects of accelerated climate change. In the United States, these include:

- Most forms of irrigated agriculture subject to surface water scarcity and groundwater depletion;
- Most orchards in areas experiencing milder winters and hotter summers;
- Nearly all grass-fed livestock producers, who have been forced by drought to seek supplemental forage or hay;
- Any gardens, fields, orchards, or pastures that are dependent upon wells or artesian springs in areas where recharge has been diminished; and
- Some forms of freshwater aquaculture, which have already been affected by salinization and water scarcity.

How Will Climate Change Hit Us? Let Me Count the Ways . . .

To be sure, the higher frequency of catastrophic weather events like those already suffered in more than 25 US and Mexican states are not the only consequences of climate change that will directly affect our food-producing capacity. In study after study—many of them summarized in 2008 by a US Climate Change Science Program submitted to Congress[4]—the following climatic impacts of food production and distribution have been documented:

- More frequent extreme heat in summer;
- Milder winters, resulting in a lower number of seasonal "chill hours," which are required by many fruits and nuts before they can break dormancy and blossom in the spring;
- Greater water scarcity;
- Longer frost-free seasons, requiring that crops receive extended irrigation;
- Greater degradation or salinization of soils;
- Greater costs of getting agricultural inputs from distant suppliers or food products to distant markets; and
- Spread of crop diseases, pests, and weeds, as well as livestock diseases, pests, and parasites previously unrecorded in certain areas.

But perhaps the key question is not about what has already happened. It is about how future climatic shift and disruption will affect *you*, and what you can do about it. It is my hope that this book will help you augment your own understanding of what may happen to gardens, farms, and ranches in your locale, and how that will influence what food shows up on your table. Ask yourself this simple question: *How may climate change affect my capacity to grow and access nutritious, affordable food over the rest of my lifetime?*

Of course, there are many possible answers to this question, and each may have a different scale of geographic specificity and time span. Let's look at just a few projections that appear to have the weight of current scientific veracity behind them, while acknowledging that there remains considerable uncertainty associated with any climatic prediction:

- In some of the United States, farmers and ranchers may anticipate a 5°F (3°C) increase in average annual temperatures by 2100, with summer highs reaching new extremes while frost-free seasons extend several weeks longer. (Perhaps that's good news *and* bad news, depending upon where you produce food!)
- Many locales are likely to receive less precipitation than their running averages over the last century, but more universally, there will be greater rainfall uncertainty and perhaps a higher frequency of catastrophic events such as droughts and floods.
- While some biophysical models predict the higher productivity of certain crop species due to atmospheric carbon enrichment and other factors, other ecological impacts like

This ancient irrigation system in Oman is still fed by a traditional *falaj* rainwater catchment system, but nearby groundwater pumping and the resulting salinization have already led to the abandonment of modern irrigation agriculture after just three decades.

the spread of pests, weeds, and diseases may negate or at least sporadically disrupt this yield potential.

- Because the growing season may extend longer in certain landscapes, there may be a greater need for irrigating crops for longer periods, or for providing irrigated pasturage for livestock.

- There are likely to be accelerated losses of carbon in soil organic matter from those farmland soils that currently support annual crops, in part because of extended growing seasons and higher heat loads.

- If summer storms become more intense in some locales, they may trigger greater soil erosion by winds and floods.

- Some aquifers may suffer declining groundwater levels due to the higher crop irrigation demands not being met by scarce surface water resources. More springs are likely to dry up in drought-stricken and aquifer-depleted areas, restricting their continued use by farmers and ranchers.

- There probably will be disruptions in the myriad relationships between crop plants and their pollinators. The disruption of these ecological interactions may reduce the yields and quality of certain insect-pollinated fruit, nut, and vegetable crops. In particular, the timing of emergence or arrival of pollinators may get out of synchrony with the flowering times of particular short-cycle crops.

- Longer and more intense fire seasons are likely to affect rangeland productivity and water quality generated from headwaters in wilderness areas.
- The frequency, duration, and severity of insect and disease outbreaks may shift with longer, warmer growing seasons and milder winters.
- Where climatic conditions become harsher or include more frequent catastrophic events, the infrastructure supporting agriculture and food transport is likely to deteriorate faster than it can be restored. In particular, irrigation canals, dams, bridges, roads, and railroads will require greater maintenance.

Of course, these trends will not occur in isolation. Perverse synergies (or positive feedback loops) may emerge as one factor influences another. The compounded consequences of climate change are likely to cripple the current means our country relies upon to feed itself through

In many of the world's deserts, accelerated groundwater depletion and salinization in the face of climate change is driving the abandonment of orchards, fields, and vineyards, like this one that once grew table grapes near the Gulf of California in northern Mexico.

conventional agriculture and livestock production. Such disruptions of our food-producing capacity will certainly not be limited to semi-arid and arid regions alone, but become more pervasive in wetter temperate and tropical landscapes as well. Perhaps more than other traditional forms of food production, from biodynamic farming to traditional Latino dooryard gardening, *agribusiness as usual will be stripped down and restructured* by somewhat novel, profoundly challenging climatic conditions. Its current emphasis on genetic uniformity, monocultural plantings, and high inputs of fossil fuels and water may be its own undoing.

And yet, the notion of the new normal (non-stationarity) does *not* imply that we need to dismiss everything that has worked in the past for growing food. It does not preclude us from drawing upon the historically effective practices used on traditional desert farms as one of many means to adapt to climate change. In fact, one goal of this book is to carefully evaluate time-tried traditional practices from desert farmers around the world to see if they may have broader applicability. My old friend Julio Betancourt, a senior scientist with the US Geological Survey, puts it this way:

> There [has been] some discussion about whether history is still relevant in a non-stationary world. It most certainly is. Knowledge about ecological responses to past environmental variability is essential to shed light on the range of potential responses to future conditions. Historical trajectories of pattern and processes in ecosystems and landscapes should continue to be integrated into projections of future climate, land use, and invasive species spread . . . In a non-stationary world, continuity of observations will be of even greater value. Stationarity may be dead, but history is alive and well.[5]

Parables

Over half of my life, I have gained considerable comfort and insight from the wisdom of the desert offered in the parables and proverbs of the monks, nuns, and hermits who left the comforts of civilization for the deserts of the Sceti and Nitria in Egypt.[6] Even today, a thriving rural economy based on desert-adapted sustainable agriculture exists in Wadi al-Natrun in the heart of the Sahara Desert in North Africa. It is focused around monasteries rebuilt on the sites of ruins left behind by the first Desert Fathers in the fourth century AD. Whenever I have visited Wadi al-Natrun, I have come away elated by the fact that an ancient *desert-adapted* contemplative tradition has been revived there. More specifically, it has guided the emergence of an alternative food production and distribution network that now feeds hundreds of monks, as well as thousands of the food-insecure poor dwelling in Cairo and the upper reaches of the Aswan region.

It is clear that the desert hermits who historically lived near Wadi al-Natrun—the Dry River of Nitrate Salts—observed significant changes in climate and land conditions even in their day, and that they anticipated more to come. In a rather ominous story that is perhaps more pertinent today than it may have seemed 14 centuries ago, Abba Macarius the Great explains to the monks how to read the signs that will tell them whether their own desert is being devastated:

> When you realize that someone can come and build on dry ground right where you knew we once had a tiny marshland around a spring, know that the devastation of the Desert of Scetis is near. When you see a different kind of tree encroaching on the desert itself, know that dramatic changes have already arrived at our door. And when you see young children arriving as orphans and refugees in our midst, take up your sheepskin bed mats and flee ...

I have been particularly moved by this story, because the monk who has twice shown me around Wadi al-Natrun is also named Macarius. When he first came to dwell in the desert, he adopted the name of that ancient camel-driver-turned-hermit in an act of inspiration. But monks like Macarius the Great—those who initially established the contemplative tradition in Wadi al-Natrun between AD 285 and 330—surely did not expect everyone to be able to live with the austerity of a truly desert-adapted life. Their admonitions remind us that not all of us may be mentally or emotionally equipped—at least at present—to patiently practice the kinds of desert survival strategies that got them through, but we can all metaphorically learn from their tradition nonetheless. One story told by the Coptic Christian monks of Wadi al-Natrun today offers all of us a way to learn from their desert experience:

> There was once in our monastery a newcomer called Abba Gelasius who was often stricken by fear with just the mere thought of going to the desert to live the hermit's life. Disappointed with himself, he

finally worked up the nerve to tell his brothers that he needed to do a trial run of desert living, but wanted to first practice the desert hermit's discipline within the confines of the monastery's walls. So he told the other monks not to bother him for a while if they saw him walking around this cloistered space in silence.

Abba Gelasius then began his mini-pilgrimage through the small patch of desert held within the monastery's walls. Although the other monks did not speak directly to him, he soon realized that his head was full of his own talk and fear. As he kept on talking to himself in his head, he pretended that he had actually ventured out into the surrounding desert.

He would proclaim to himself, *"He who walks in the desert does not partake of oven-baked bread, but instead relies on wild desert herbs. So when one gets hungry and weary in the absence of bread, one must feel free to eat the sparsely-populated vegetables on the desert floor."*

Abba Gelasius then ate the desert herbs and vegetables until he was no longer hungry, but his weariness persisted. And so he spoke to himself again, saying:

"He who becomes one with the desert has no need to lie in a bed, but should give up the fear of sleeping in the open air on the barren ground."

So he lay down and slept in the bare spaces within the monastery walls, as if he were truly out in the desert. He continued on his imaginary pilgrimage into the desert for three days more, constantly circling around within the walls of the monastery, eating desert chicory leaves wherever he found them and then sleeping in the open air whenever weariness took over.

Finally, he realized what the matter was, and got to the root of what had troubled him and kept him from being a true desert hermit. It was not that every faithful soul should go out into the true desert to eke out a living, but that they should find a desert-like space and overcome their fear of it wherever they may already be dwelling. If he was not able to practice the way of the desert in his very own home place, there was no reason for him to venture out past the limits of his own capacities. He need not wander away into the Desert of the Sceti to become an ascetic. Instead, he needed to learn patience and tenacity in his everyday environs, humbly adapting to the place where his own lot had already been cast.

Perhaps Gelasius eventually learned a kind of patience exhibited by the most diligent of the Desert Fathers; that patience may also be the key trait that we need to cultivate in the face of climatic uncertainty. To glimpse at the kind of patience our own journeys may require, let us turn to a parable about Abba John Colobus,[7] a dwarf who as an 18-year-old had ventured out to live in the harshest reaches of the Sahara around AD 357:

The monks recall the time when Abba John Colobos, a dwarf, withdrew from their company to go to live with an old Holy Man far out in the Desert of the Scetis. This Holy Man was none other than Abba Ammoes, who had originally come from Thebes. One day after John the Short had arrived at the cave of Abba Ammoes, the Holy Man took a leafless, barren piece of dry wood lying on the desert floor, and demonstrated how to carefully plant it in the ground. He then proclaimed, *"My little Dwarf, please irrigate this tree every day with a pitcher of water until it bears fruit."*

The Holy Man said no more than that before he abruptly left Abba John the Short to fend for himself, but both knew that the closest water source was a long way from them. Abba Ammoes retired into the shade of the cave and began to weave a basket. For the Dwarf to reach and fetch enough water to irrigate the piece of dry wood planted in the ground, he had to journey to a spring far away from their cave. Worse yet, because his legs were so short from being stunted at birth, the Dwarf had to leave the cave in the cool of the evening, walk half the night, dip a giant urn into the spring water until it was filled, and begin to walk back home at dawn with the burden of the urn upon his shoulders. He would arrive back at the planting halfway through the following morning, for the weight of a water urn filled to the brim inevitably slowed him down. Nevertheless, John diligently went to fetch some water day after day, and obediently irrigated the dried-up piece of wood that others dismissed as being the pitiful remnants of a fossilized tree.

At the end of three years, John the Dwarf went running to Abba Ammoes to tell him that he had noticed something new at the site of their planting. The Holy Man left his cave and came forth to follow John out to the place where his instruction had begun years before. They saw that the old piece of wood was not dead at all, let alone fossilized; it had merely been dormant. Finally, it had received so much moisture and care from John the Short that it had come alive once more, sprouting leaves while little fig-like buds formed on new branches. After several more irrigations, John the Short and his teacher Abba Ammoes watched as the figs miraculously ripened into full-sized fruits. The Holy Man then helped John the Short climb up just high enough into the tree to pick its delicious fruits and hand them down. They filled the wicker basket that the Holy Man had been weaving over the intervening years.

When the dwarf descended to the ground from the canopy of the tree, the Holy Man beckoned him to help carry the basket filled with ripe figs over to the monastery from which they had both departed years before. As they entered the monastery grounds, Abba Ammoes shouted for all the monks to join them in the chapel, where he poured the figs out on the altar. The Holy Man then said to their brethren,

"Behold these ripe and luscious fruits, for they are the bread of patience and obedience that John the Short has carefully leavened and tenaciously kneaded since you last saw him amongst you. His attentiveness and diligence are what have brought us the nourishment that can now be shared among us. Take, and eat from this living bread made possible by John the Short's dedication to watering what most of you might have mistakenly taken for the dead!"

I often think of this parable while I am out building water-harvesting structures on my land on days when there is not a rain cloud in the sky, nor any real soil moisture on the surface of the land that I am reshaping in my attempts to capture and retain more humidity. How do we train ourselves not to grow so despondent that we fail to do work that might allow us all to reap benefits many months or years from now?

Principles and Premises

Given the climatic trends being documented by scientists and the moral warnings being offered by ethicists and spiritual leaders, how do we respond? How do we get a grip on the climate change that we are likely to experience over the rest of over lifetimes, and how to we adapt our production and use of food to new realities? To begin with, what are some ways that we can anticipate climate's potential effects on our own capacity to produce food? Here are some preliminary considerations:

This simple shade structure at Robert Duncan's orchard near Victoria, British Columbia, protects citrus trees from climatic uncertainty.

- Anticipate the unexpected: Catastrophic weather events may be arriving at times and with intensities that we have not been sufficiently able to predict.
- Assume that temperature maximums will be higher and durations of hot spells longer, placing unusual stresses on crops and livestock.
- Anticipate higher evaporation rates from the surfaces of the soil and various bodies of water, and higher transpiration from the leaves of crops. Our crops and livestock will demand more water to produce yields comparable to those of the past.
- Get ready for longer frost-free seasons, which may create greater water demands by our crops and livestock and increase our reliance on supplemental irrigation (hopefully, with rainwater and graywater).
- Anticipate more variability in rainfall from year to year, with some annual totals reaching record lows, while droughts may endure longer than in the past.
- Be assured that at least some rivers and reservoirs now used for irrigation will dry up, and that we will see slower rates of aquifer replenishment and more rapid depletion of groundwater due to overpumping.

- Anticipate that ranchers having more difficulty accomplishing year-round grass-fed livestock production and farmers in semi-arid areas may have to abandon dryland farming of forages or food.
- Be alert to the introduction of new insect pests and insect-transmitted diseases, with higher levels of insect damage and viral infestation becoming evident on crops that are stressed by drought.
- Stay alert to the arrival of new invasive weeds, and to potentially heightened crop damage by vertebrate pests.
- Consider the possibility of disrupted **pollinator services** as bees get out of synchrony with the flowering times of the plants that would otherwise provide them with pollen and nectar.
- Watch out for deteriorating road, railroad, and bridge infrastructure, which will make it more costly to move food long distances.
- Be prepared to absorb higher fossil fuel and fossil groundwater pumping costs.
- Seek out local options for inputs to your food production as your access to extra-local nutrients, fertilizers, and materials becomes disrupted or restricted.

No Sniveling or Hand Wringing Allowed: Here's What You Can Do!

While some of the changes in our food system to be brought on by climate change may seem inevitable and inexorable, you are not a victim! There are many ways that you can begin to adapt your food production and consumption to not only reduce the effects of climate change, but control some of its causes.

Much of the rest of this book will focus on the positive ways you can deal with climate change with your food-producing practices and with your fork.[8] While other chapters will detail specific strategies for adapting our food production to a changing climate, here are a few of the principles to guide your actions, and those of your allies, in growing gardens, fields, or orchards, or in raising farm animals:

- Reduce the amount of fossil fuel you use in pumping water, tilling the soil, carrying inputs like compost to your food production area, and transporting produce to market or back to your own table.
- Redesign your food production, harvesting, processing, storage, and consumption to be more cohesive, with shorter distances between your links in the food chain.

- Retrofit your foodscape to be better buffered from floods, winds, and catastrophic weather by forming rainwater catchments, vegetated field margins (**fredges**), swales, and water surplus sinks to reduce the force of these stressors.
- Increase the amount of living ground cover on your land in ways that sequester carbon and increase soil organic matter, holding as much root mass and soil in place as possible.
- Explicitly plant more deeply rooted perennial trees, shrubs, vines, and succulents with **biochar**, earthworms, effective microbes, and **mycorrhizal** fungi in each hole.
- Develop **nurse plant guilds** with protective "nurse plant" perennials that provide shade, soil nutrients, and mycorrhizae to **understory** plants that do not require full sunlight to grow.
- Use gravity-fed and solar-pumped rainwater stored either in the soil, or in tanks adjacent to fields, orchards, and gardens so that they may irrigate crops on demand.
- Opt for heat- and drought-tolerant food crop varieties that are less water-consumptive, then further select and adapt them to changing weather conditions through on-farm breeding or mass selection.
- Select varieties of perennial fruits and nuts with chill requirements of 200 to 250 fewer hours than the winter chill accumulation needs of the fruit trees currently dominating in your area, in the hope that these will continue to be productive for many more years as average winter temperatures rise.
- Plant annual crops with the onset of seasonal rains, letting them wind down their production near the end of the rainy season rather than extending their irrigation well past the natural period of accumulated moisture.
- Remember that half of your ecological **foodprint** (and half the waste) is generated between the time your food is harvested from the land and the moment when it reaches your table.

Planning and Practice:

Establishing What Your Baseline Conditions Are

To truly get a grip on climate change, you must become attentive to the place-based causes and effects in your own foodscape. No outside expert can really do that work for you. Since the advent of the new normal under which there is rapid change, conventional guides like the USDA Plant Hardiness Zone Map have become less helpful. And so it is critical that you begin to think of ways that you can gradually characterize your current climatic, hydrological, and ecological context. Try to monitor how rapidly it may be changing, and refine your responses to fit the new conditions that are emerging in your area. Here are some warm-up exercises you may want to consider:

- Obtain from your local library or from online sources the longest set of weather data you can find taken on or near your locality, and begin to chart changes that have already occurred over the last quarter to half century. In particular, chart the year-to-year differences in length of the frost-free growing season, in the amount of seasonal rainfall (using solstice and equinox dates to mark the shifts within seasons), in the date of onset of warm-season rains of a half inch or more, in the number of days when temperature highs reached over 100°F (38°C), in the number of days temperature lows reached between 32 and 45°F (0–7°C), and in the number of days where temperatures fell below freezing.
- Look at these charts, and determine when (or if!) weather shifts began to accelerate in your area. Are they unidirectional trends, or do they suggest that your local climate is oscillating more wildly? What kind of climate would you project that your place might experience in 20 years, given current trends?
- Talk to two or three old-timers who live within 10 miles of you. What are their opinions of how much the weather has shifted? What catastrophic events (floods, droughts) do they remember, and in which years did they occur? Do they perceive that the frequency of such events is getting greater? What changes in water or vegetation and wildlife have they witnessed?
- Go online and see if climatologists have charted changes in maximum daily or annual temperatures, maximum summer temperatures, or changes in winter chill hours and growing season lengths for your locality, county, or watershed. If so, what do these trends suggest to you relative to similar charts for North America or the earth as a whole? Why might your local rates of change appear to be occurring faster or slower than what you see published for larger landmasses?
- Where does the water you use in your garden, farm, orchard, or ranch come from? Chart the depth of pumping levels in wells, and

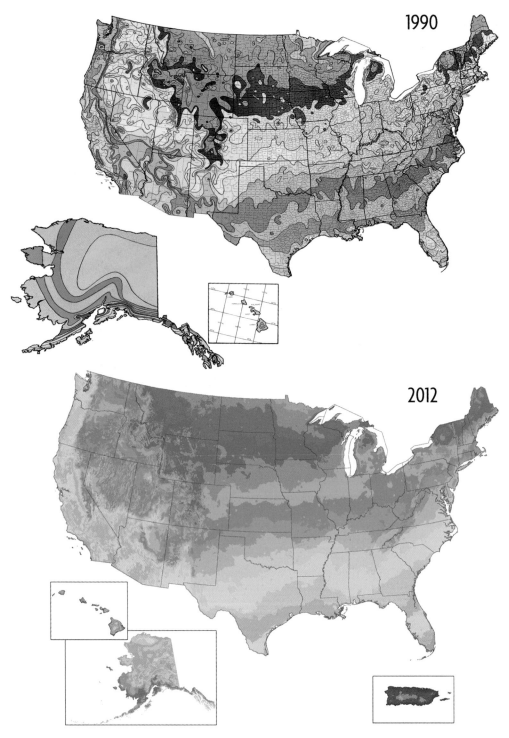

A comparison of USDA Plant Hardiness Zone maps from 1990 and 2012 showing recent effects of climate change on their distribution SOURCE: HTTP://PLANTHARDINESS.ARS.USDA.GOV/PHZMWEB/IMAGES/USZONEMAP.JPG; HTTP://PLANTHARDINESS.ARS.USDA.GOV/PHZMWEB/

the volume of reservoirs in your area over time. Is there any evidence of groundwater or surface water depletion?

- Estimate the amount of water that could be harvested from your property by multiplying your average annual rainfall by the infiltration/shedding rate for each kind of surface on your property (clay soil, asphalt, concrete, metal roofing, et cetera) times the square feet of that surface type on your land. How much water could you potentially harvest? How much holding capacity do you currently have in terms of rain barrels, catchment tanks, cisterns, and ponds? What are the most cost-effective means to increase your rainwater-holding capacity?

- Purchase or make two sets of flags—one green for "sustainable," one red or orange for "contributing to climate change"—and walk around your property or your larger foodscape, placing them next to particular features that contribute to or detract from your food-producing capacity. Does your gas-guzzling pickup truck need to be retired? Do you have an opportunity to replace an electric- or diesel-powered well pump with a solar- or wind-generated one? Is there a nick-point for gully erosion on your land that is taking out trees? Have you recently planted an orchard that has the potential of sequestering more carbon than the garden of annual crops that preceded it? Once you have all the flags planted next to particular features, map them as the first step toward developing a site plan of what needs to be phased out, retrofitted, encouraged, or newly designed on your land to reduce your ecological footprint.

- Make an inventory of all the inputs to food production that you currently obtain from beyond your home place, whether it is water diverted through canals from a distant river, pumps, pruning shears, rototillers, compost, soil mineral amendments like sulfur, pollinators, fossil fuel, or diverted river water. Next to each input and its presumed place of origin, propose how you might obtain its equivalent from sites closer to your own, or at least how you might obtain them with lower transportation costs involved in their delivery.

- Take a box of crayons and a large sheet of butcher paper, and use them to reimagine the food-producing capacity of your land. Re-dream what features you would ideally like to see in your foodscape, and picture it as either a map or drawing of that scene. Then, in the margins, place arrows showing all the beneficial connections your particular foodscape shares with the larger foodshed and watershed within which you are nested.

• Come up with a time frame over which you would like to implement each change that is within your economic means to make. Finally, list potential collaborators, micro-investors, nonprofits, or government agencies that might assist you with accomplishing the other goals that are currently beyond what your own pocketbook might allow.

Now that you can see some obvious ways to get a grip on climate change, let's try to imagine some ways to redesign your foodscape to weather the storm and produce diverse, nutritious food while the tides of change ebb and flow.

Chapter Two

Seeking Inspiration and Solutions

*from the Time-Tried Strategies
Found in the World's Deserts*

✿

Warm-Up

The challenges facing any form of food production in a time of rapid climate change can seem daunting. Indeed, they would be absolutely impossible to deal with if there were not already time-tried strategies to draw upon as means to adapt to a hotter, drier, more uncertain world.

In this way, we are fortunate: We are not beginning at the 11th hour with a disarmingly blank slate. There are many ways to adapt to drought, heat, and wilder, less predictable ecological oscillations; we can successfully adapt because they have, in fact, already been "field-tested" by nature's diverse organisms and ecosystems and by myriad human cultures. Many species—including humans—have subsisted on the 10 to 17 million square miles of arid and semi-arid landscapes on this planet over the last 10,000 years. During most of those millennia, our climate has been relatively stable for food production relative to what we may be facing over the next century; nevertheless, it has been theorized that what triggered the very origin of agriculture on several continents 8,000 to 10,000 years ago was warming and drying trends, which forced the growing human population to innovate in order to achieve greater food security. As we now enter the new normal of even greater climatic uncertainty, we may have to scale up the most promising adaptations that desert dwellers have improvised over the last several centuries to achieve greater resilience in our food systems as a whole.

In short, emulating and refining these adaptations to serve as a foundation for our own may help us secure food in the face of climate change.

At the least, they will give us much "food for thought" as we wander into a world that is quite different from the one we have lived in most of our lives.

To access, understand, and fit these modifications to our own place, we will inevitably need to tune in to a couple of vast networks of valuable information: We must remember or relearn *how nature works* to reduce stress in plants, animals, or arid ecosystems. Then we must study traditional ecological knowledge, designs, behaviors, and practices of indigenous and immigrant cultures that have long foraged, herded, hunted, or farmed in desert landscapes.

❀

Parable

In the mid-1970s, I began desert fieldwork with my closest friend, Tom Sheridan—who has since been internationally recognized as the leading political ecologist and agrarian anthropologist of the US–Mexican borderlands. But in those days, simply making field observations of desert farming and ranching was what united our pursuits. We set out together on every back road we could find, hoping to see what traditional practices Sonoran Desert farmers or ranchers still used to deal with water scarcity. It seemed to us that limited water access resulted from both the tremendous variability in rainfall that is so common in desert regions as well as the "political ecology" that allowed some desert dwellers to control and use more of this scarce resource than their neighbors.

We would drive south across the border at Nogales crossing, and within an hour's drive into Mexico we would veer off the highway onto dirt roads, dry riverbeds, and cattle trails. Far beyond the end of the pavement, we listened to the farmers, ranchers, and orchard-keepers who lived along narrow ribbons of greenery supported by the meager surface water available. Their lush farms and gardens were sparsely distributed, but scattered among the hundreds of thousands of miles of slickrock, creosote flats, cactus forests, dry *playas*, and deep canyons where agriculture, upon first glance, seemed all but impossible. Tom and I sought to learn: How did these people make food production work despite the odds stacked against them?

To gain a deeper understanding of how desert food producers dealt with the challenges just beyond their doorstep, Tom and I spent considerable time in the centuries-old farming and ranching village of Cucurpe, Sonora. There, along an intermittent stream that trickled out of canyons and springs, Spanish and Indian land and water management traditions had hybridized over three centuries.[1] It was there where we first noticed that much of the streamside vegetation alongside farms was technically not "wild." Rather, farmers had planted thousands of live cottonwoods and willow stakes in the moist sand of the streambeds adjacent to their fields, pastures, and orchards.[2]

We gradually were able to discern why Sonoran farmers made such an effort to maintain tree species adjacent to their fields of annual crops. One growing season, the farmers told us, they might be plagued by lingering drought, while the next their floodplain fields might be inundated for several days in the aftermath of a hurricane fringe storm. The trees, they claimed, usually stood their ground and buffered their fields from the ravages of both drought and floods.

But how, we wondered, did the farmers use these trees to survive such extremes and make good use of whatever water nature gave them?

I did not fully comprehend it at the time, but the Mexican and Native American farmers we met in places like Cucurpe were intensely interested in the desert's natural processes. These farmers took their cues from natural structures, processes, or behaviors that they saw in the desert world, and applied them to their stewardship of food-producing lands.

For instance, some looked for the leafing-out of acacias and mesquites to cue them that the last winter/spring freeze had passed. Others watched the placement of the sun on the horizon at dawn in conjunction with two summer lunar cycles to bracket the period within which the planting of warm-season crops would lead to a successful harvest.

Perhaps these farmers made the most direct and effective application of "designing with nature" by the way they designed their streamside fencerows to emulate what they saw working in nature. As I mentioned a few moments ago, by mimicking the buffering capacity of the native gallery forests situated along desert streams, they designed their living fencerows or streamside hedgerow networks to fulfill many of the same functions found in those riparian habitats.

Using a vocabulary quite unlike that of ecologists, these Sonoran Desert farmers had historically recognized and enhanced several **ecosystem services** provided by managed living fences that mimicked adjacent wild riparian vegetation. The farmers realized that by intentionally planting curvilinear corridors of cottonwoods and willows, they could:

- Protect their fields from ravaging floods;
- Capture both nutrient-rich water and organic matter that could be used to enhance fertility in their field soils;

A *fredge*, or living fencerow, shades an adjacent field, cooling the crop canopy while providing other benefits as well. DRAWING BY PAUL MIROCHA.

- Offer early-morning and late-afternoon shade and coolness to reduce stress on summer plantings;
- Protect their plantings from browsing or trampling damage by free-ranging cattle, horses, or wildlife; and
- Provide wood both for construction use and for kindling through the pruning or **coppicing** of lower branches on the living fencerows (*cercos vivos*) that we will hereafter call *fredges* (see the sidebar).

We're now aware from numerous field surveys that fredges sequester carbon, reduce crop losses by reducing the speed and turbulence of winds blowing across fields, slow evaporation and soil desiccation, halt soil erosion and nutrient runoff, serve as corridors for migratory pollinators and other beneficial wildlife, and host insectivorous birds—which, in turn, dampen pest population fluctuations.[3]

While scientists have documented at least 20 other ecologically and economically important functions that fredges offer to a healthy **farmscape**,

What's in a Name?
Hedge, Fedge, Living Fencerow, Fredge . . .

American English may be the most awkward language in the world for offering a precise typology to describe various forms of vegetated field margins such as hedges, windbreaks, shelterbelts, and fencerows. We simply don't have a generic term as melodic as the ancient, pre-Columbian phrase in the Nahuatl language, *tlaquaxochquetza*, which Aztecs might have roughly translated as "an erected border of woody plants serving as a field boundary."[4] Neither do we have the easy but overlapping categories found in Latin American Spanish such as *cercos vivos* living fences, *cercos de tejido* woven fences, or *cercos de rama* woven brush fences. The term *hedgerow* as it is used in the British Isles seems so formal and historically rooted in the agrarian landscapes of the Old World that few Americans call their vegetated field margins a hedgerow; across the pond from the Brits, the term is usually reserved for some ornamental plants up against the wall of a house or garage in a suburban yard. And so, agricultural geographers began to use the rather clunky term *living fencerow* to distinguish the linear plantings of live stakes of ocotillos, prickly pears, bamboo, carrizo, cottonwood, or willow from the more rotund, three-dimensional English versions.

A tentative truce between British and American agricultural geographers over terminology was brokered when a young British landscape designer trained at Oxford tried to popularize the term *fedge* in her book *High-Impact, Low-Carbon Gardening*,[5] which was simultaneously released on both sides of the pond. And yet the term has not gained much currency, so I will propose another, hopefully more memorable one. I suggest that we rally behind another syllogism, the *fredge*, which takes its *f* and *r* from *fence* and *row*, and the rest of its letters from *hedge* and *edge*. It may have another advantage in the United States, since every red-blooded American male already knows how to "head for the fredge" when it's time to get something to eat!

Fredges, shelterbelts, and windbreaks are essential to water conservation and crop production in windswept Tajikistan.

This living fencerow or fredge not only protects field soil banks from erosion during catastrophic storms, but captures nutrients to regenerate the field's fertility and productivity.

they are only now beginning to assess how such human-tended corridors of streamside vegetation can mitigate against the effects of climate change.[6] It seems to me that science is belatedly recognizing the value of these "green veins." Fredges and other forms of vegetated field margins clearly date back to pre-Columbian times and were historically, if not prehistorically, commonplace throughout Mexico and the southwestern United States.[7]

Ironically, as climate change has accelerated over the last half century, government programs have forced many farmers to rip out these multifunctional fredges from field margins in the name of "water conservation" or "food safety." Other kinds of vegetated field margins have been eliminated to provide more room for navigating tractors around the edges of row crops, or to allow planting of cotton or alfalfa from one property line or fence all the way to the next. Sadly, the Rio San Miguel and Rio Sonora of eastern Sonora, Mexico, are among the very few watersheds in the deserts of northern Mexico where vegetated field margins that buffer farmers from climatic disruptions not only persist,

A living hedgerow or fredge protects field edges and captures nutrient-rich organic detritus from upstream for soil building and enhancement in eastern Sonora, Mexico.

but are still being constructed today.[8] Similar fredges, once common-place north of the border in Arizona, have virtually vanished from these agrarian landscapes, thus ensuring a great vulnerability in the face of floods and droughts.

But how did Sonoran farmers learn to "design with nature" in the first place? Inspired by the ecological functions offered by the wild willows and cottonwoods, some innovative farmer (whose name is now lost to us) realized that he could propagate these same trees alongside his fields with vegetative cuttings and gain the same functions. In essence, the Sonoran farmer had learned to "think like a desert riparian forest" in order to buffer his fields, pastures, or orchards from climatic extremes.

Perhaps this farmer first observed that floodwaters could be a bless-ing, not just a curse. He recognized that the *agua puerca*—silt-laden floodwaters—carried valuable nutrients (*abono del río*) when stream flows surged above the stream channel and inundated much of the floodplain after the intense monsoons of summer and fall. I tangibly learned about the value of these floodwaters by wading out waist-deep in inundated arroyos, bottling some of the water and streambank during raging storms, and analyzing the debris they left behind in the desert as they carried enormous quantities of nutrient-rich compost and mulch to places where such nutrients might otherwise be hard to come by. The murky floodwaters were filled with rich soil and the half-composted leaves and stems of nitrogen-fixing legumes such as acacias and mesquite, along with jackrabbit and packrat droppings and cow and horse manure.

The stronger the flood, the more likely it would also haul along branches and uproot trunks of trees as well. Any farmer looking at the flotsam and jetsam left by a desert flood would quickly realize that he could make good use of a fresh flush of soil and compost in his fields, if he could just find a way to filter out unneeded trunks, large branches, rocks, and sand.

The next ecological insight the Sonoran farmers came upon became key to their designing highly functioning fredges. They noticed that the good soil and compost tended to accumulate just downstream from cottonwood and willow trunks, while other materials—branches, trunks, tires, and boards—were "screened out." Further, where such branches became stacked up between two standing tree trunks, they began to serve as a filter or sieve that let the silt and compost through while holding back larger debris.

Farmer Beto Cruz once told Tom and me that in a well-placed fredge, "The trees and woven branches accept the floodwater and make it tame."[9] By planting live cuttings of cottonwoods and willows a few yards apart from one another and weaving branches between them along or near the edges of their fields, farmers could grow a living fencerow that held their field edge in place during floods, replenished the field's nutrients, and sieved out the debris. As the floodwaters hit the woven fence, the reduced velocity of the waters forces them to dump their bed loads, spreading silt and compost downstream from the fence over the surface of the field.

Furthermore, the best designers of these living **silt traps** have found ways to gain maximum benefits from their plantings by aligning them in curvilinear rows that echo the natural meanders of desert rivers, and situating them 10 to 15 feet out from the more elevated field edge on the floodplain. The most talented practitioners who use fredges to artfully reclaim field soils—such as our late mentor Don Beto Cruz of Cucurpe—have not only expanded the area of arable land available for planting, but renewed soil fertility without ever having to purchase fertilizers or other commercial soil amendments. Don Beto let the floodwaters carry and deposit compost in his field, rather than expending the energy of hauling and dumping. As his relative Ana Cruz once told us, "The fencerows actually form the [arable] land." Essentially, Beto, Ana, and their neighbors could "grow" the fields out into the more gravelly streambed by encouraging the deposition of the soil carried by floodwaters. This soil deposition filled up the space between the current field edge and the desired edge immediately behind the fredge. Don Beto claimed that floodplain fields managed by his family would stay fertile for at least five years after a major flood. My studies confirmed that his soil had levels of organic matter and moisture-holding capacity comparable to the best midwestern and southern farms along the Mississippi River.

While fredges do indeed buffer Sonoran farmers from the kinds of perturbations wrought by climate change, my purpose here is not merely to encourage each of you to go out and find a place to plant a fredge on your own land. Rather, my goal is much broader: I'd like to underscore the value of **biomimicry** and related design principles that encourage you to "think like a desert dweller would" in the face of climate change, whether that "dweller" is a giant cactus, a horned lizard, a camel, or a nomadic Bedouin sheepherder.

❁

Principles and Premises

There are design principles and operating instructions embedded in nature and culture that we can use to ecologically redesign farming and food systems. One way to activate these principles and instructions is by becoming conversant with the methodologies of the emerging discipline of biomimetics. First codified in 1997 by biologist Janine Benyus in her groundbreaking book *Biomimicry: Innovation Inspired by Nature*,[10] **biomimetics** is a creative process through which the earth's flora, fauna, and microbiota are emulated to design a particular human-oriented product or process. Although Benyus explicitly identifies innovations inspired by natural forms (morphological and anatomical structures), natural processes (biochemical pathways and behaviors), and their roles in natural ecosystems (ecological services), she begins her classic work by giving a nod to the wellspring of traditional ecological knowledge from indigenous cultures. For our purposes here, we might consider three levels of innovations inspired by deserts and their place-based cultures to serve as points of departure when designing resilient, adapted agro-ecosystems that can survive climate change:

1. Innovations derived from classical biomimicry, where the forms and processes found in individual desert-adapted organisms or their populations are used as inspirations;
2. Innovations derived from eco-mimicry, where the functions and structures of symbiotic relationships, guilds, communities, and eco-systems in arid regions are used as inspirations; and
3. Innovations derived from ethno-mimicry, where the traditional ecological knowledge of desert-dwelling farmers, foragers, orchard-keepers, hunters, herders, and ranchers is used to inspire new emulations (since they cannot and should not be copycatted).

All three mimetic processes can encourage a natural systems agriculture that will be better adapted to climatic changes.

Early on, Benyus recognized that grassland ecologist Wes Jackson and his staff at the Land Institute were actually using the prairie as a model for perennial agriculture that intercropped cereal grains, legumes, and oilseeds without any need for recurrent soil tillage. Wes Jackson's four books and thousands of barnstorming lectures have galvanized many prairie state residents around this new kind of **perennial polyculture** that builds rather than depletes soils, as well as their carbon- and water-holding capacity. Benyus deftly extended his analogy, realizing that novel forms of food production can be based on any natural ecosystems if ecologists, soil scientists, plant breeders, environmental historians, and others were brought together into design teams where they explored how to emulate the ways those natural ecosystems work (see below). Together with architects, systems designers, farmers, nutritionists, chefs, and food supply chain managers, they might ask:

- What native food species grow here that can stand the test of time—or the test of climate change?
- What food-producing organisms together grow here that collectively or synergistically provide greater resilience than any one species in monoculture might do alone?
- What culturally managed species have been united to produce edible landscapes that, when guided by traditional ecological knowledge, have comparable resilience as natural ecosystems?

And so, if we are to specifically use the wisdom found in deserts and their ancient cultures to design healthy, resilient food-producing communities with small carbon foodprints, we might rephrase our operating instructions in this manner:

Biomimicry: Becoming Co-Designers of a Climate-Resilient Foodscapes

When the term *biomimicry* was first coined in the 1980s, and even when it was popularized by Janine Benyus in the late 1990s,[11] few **permaculture** activists fully understood that climate change was already having a perceptible effect on food production. Nevertheless, many of the design principles elucidated by Benyus and her colleagues at the Biomimicry Institute could play an enormous role in helping humankind adapt and diversify agriculture to weather the next century of accelerated climate change. Below in bold are the principles (as expressed by Benyus[12]), with my efforts to link them to climate-friendly food systems in italics:

- **Nature runs on sunlight.** *To be sustainable, your food system must be redesigned to have a lower carbon foodprint, being run largely on solar energy rather than on fossil fuels.*
- **Nature uses only the energy it needs.** *You will target the energy flowing into your freshly designed food system to produce and distribute food with the highest ratio of nutritionally dense calories brought to the table to number of calories expended getting it there (while maintaining a resilient infrastructure to do so).*
- **Nature fits form to function.** *The crop varieties you select will be adapted to the environment, rather than requiring you to use excessive water and fossil fuel to remake the environment to fit the crop's needs.*
- **Nature recycles everything.** *No organic materials will be left behind; you will reduce your ecological foodprint by trying to repurpose almost everything!*
- **Nature rewards cooperation.** *You'll never farm alone; watersheds, soil microbes, pollinators, and other neighbors will be there as allies farming with you!*
- **Nature banks on diversity.** *If your foodscape is designed to be rich in agricultural traditions, species, varieties, microbes, and strata, it will offer much of the resilience you need in the face of climatic disruptions.*
- **Nature demands local expertise.** *While seeking inspiration for redesigning your food system, draw upon the wisdom of local (or bioregional) cultures, species, habitats, and plant associations or guilds. Let your homeland be your mentor, your genius loci.*

- Desert dwellers make use of abundant sunlight as their primary means to fuel food production, but they also know how to deflect its heat to reduce moisture loss.
- Desert dwellers only use the water and energy they immediately need, and try to store any excess underground for periods of scarcity.
- Desert dwellers are ergonomically shaped and prone to behave in ways that reduce drought and heat stress, but often do so most effectively when clustered or "stacked" into the multiple vertical strata known as nurse plant guilds such as those in oasis habitats.
- Desert dwellers rapidly and efficiently recycle resources whenever both water and energy become available to do so.
- Most desert dwellers may be biologically programmed to compete with one another for scarce resources, but cooperation, synergy, and

symbiosis have nevertheless become the norm more than the exception within and among desert communities.

- While desert communities may appear sparsely populated and less diverse than tropical or temperate rain forests to the unacquainted eye, they offer impressive levels of diversity among soil microbes, plants, pollinators, and frugivores, but on temporal and spatial scales very different from those of forests.
- The crop species that work well in stable, temperate climates may fail under more arid conditions. Deserts have evolved not only unique arid-adapted species, but unique webs of relationships to efficiently capture sunlight, water, and fertility to produce food. They can survive and even thrive in settings where "generalists" suited to stable, more moderate conditions can't make it.
- Deserts have little tolerance for excess; they are stripped-down, elegant ecosystems that can rapidly shed any excess baggage during periods of stress. Desert shrubs often drop their leaves and reduce their root mass during extended droughts, only to rapidly regrow them following sudden bursts of rainfall.
- Desert dwellers humbly accept the inherent constraints of the desert. As I've noted, a Pulitzer Prize–winning western writer once quipped that newcomers to the desert must learn to "get over the color green." Because many of us recognize that water scarcity is a norm, not an aberration, the water-saving strategies of desert plants can inspire us to innovatively make the wisest use of the resources that are available.

I do not assume that all readers and users of this book will be residents of arid or semi-arid regions of this planet. But I sense that every farmer, rancher, or orchard-keeper at this point in time can gain something by considering and implementing some of the desert's basic "operating instructions." It is in this way that I am encouraging you—as food producers, consumers, and visionaries—to use the metaphor of desert survival as a means to go beyond biomimicry as an exploratory tool that works merely for the design of individual products or processes.

In effect, it is not enough to simply put a drought-adapting or heat-stress-reducing structure in place, for you will need to learn how to integrate its management into your daily behavior if it is to be effective over the long haul. For instance, my straw bale house has electric shutters over the windows to reduce heat loads over the summer and retain heat during winter nights, but until I learned to immediately activate them as soon as stressful weather conditions began to set in, I hardly garnered any of their benefits. In short, each of us needs to embrace the deeper ways to adapt to and survive climate change, rather than thinking it is about purchasing or constructing quick-fix gimmicks.

I am asking you to get under the skin of certain desert plants, communities, and cultural manifestations and to "think" (intuit, behave, perform) as they do. Just as Aldo Leopold asked wildlife conservationists to engage in "thinking like a mountain," my challenge to food producers is to engage in thinking like a century plant, a tepary bean, a semi-arid prairie, or a desert oasis. Let's see what this might look like.

Biomimicry 1:
Thinking Like a Century Plant

Century plant is the common English name for a succulent species in the genus *Agave*. They are desert plants, known in the Spanish-speaking world as *mescal*, *maguey*, or wild relatives of the blue agave of tequila fame. For nearly 50 years, a plant explorer and mentor of mine, Howard Scott Gentry, searched all over the Americas for agaves, and trained dozens of us to think about the roles century plants played in desert landscapes. One of the xerophiliacs (dry-climate lovers) trained by Gentry was a brilliant desert ecologist named Tony Burgess, who later played a pivotal role in the development of Biosphere Two, the constructed and enclosed ecosystem built by the Bass brothers and now managed by the University of Arizona near Oracle, Arizona. During the time that several of us apprenticed for Gentry and worked to describe the past and imagine future roles for century plants in native agriculture, Tony synthesized most scientific knowledge regarding agave adaptations to aridity. In what has become a classic work, he suggested that:

> Ultimately it would be desirable to generate a set of "rules" summarizing the adaptive responses of individual organs [found in century plants] to specific components of natural selection.[13]

Although Tony's focus was honed in the specific quest to understand how century plants evolved to tolerate hyper-arid desert conditions in the US Southwest and northwest Mexico, his work can now help us to think like a century plant.

Many century plants are rainwater harvesters in the sense that their rosettes of sword-shaped leaves foster the trickle of both rain and fog-like mist back to their roots. (By piling pebbles and small stones around the base of cultivated century plants, prehistoric farmers in Arizona actually

Agave leaf shapes drain rainfall down toward the leaf bases and roots of the plant. We can all learn to think like a century plant.

created "fog and dew collectors" that not only increased yields but also fostered the propagation of more agaves.) Like other succulents during the hottest, driest months, they can slough off much of the mass of their root hairs and thus reduce their living biomass requiring support. And in turn, they can quickly grow them back once the rainy season arrives. Both their narrow, thick leaf morphology and grayish green color allow them to reduce heat loads during hotter, drier seasons. In short, they have a low surface-to-volume ratio by virtue of their aboveground architecture, yielding minimum water loss. (Once, on a desert island in the Gulf of California, I assisted a Seri Indian elder in the harvest and pit-roasting of century plants during a dry year, and when annual rainfall had been less than 4 inches for several years: The plants had survived and thrived largely through the moisture left on their leaves by occasional coastal fogs!)

Century plants have adapted to water scarcity and heat by fixing most of the carbon they gain through photosynthesis during the night rather than during daytime hours, when both heat stress and water loss from leaves would put them more at risk. Through a metabolic pathway nick-named CAM (crassulacean acid metabolism), they can keep their leaf pores (stomata) closed during the day to reduce water loss, while still

Century plants would not be able to grow in cracks in bedrock were it not for their capacity to collect and channel rainwater off their curvilinear leaves.

photosynthesizing. Then they open their pores during the night, when it is cooler and water loss through transpiration from their leaves is less, and they produce a four-carbon compound, which they can store deep in their tissue. Only as seedlings or after extended periods of high rainfall do they open the stomatal pores during the daytime.

If equal parts water were applied to an annual crop (such as corn, sorghum, soy, or wheat) and a field of giant century plants used for making fermented *pulque* beverages, the latter would easily surpass the cereals and legumes in productivity per acre. In fact, they can generate twice the dry biomass per area as hybrid poplars grown for pulp, and they produce three times as much sugar as sugarcane yields in the same area of planting. When water is available in the desert, they use it 10 times more efficiently than cereal crops. Already adapted to the arid and semi-arid conditions currently found on one-fifth of the planet's land surface, they can easily withstand drought, heat spells, and rising carbon dioxide levels in the atmosphere. Within the 200 million acres of their native range in Mexico, agaves can annually produce as much as 5,600 million tons of edible or drinkable biomass on far less water than almost any other crop, with the possible exception of prickly pear cactus.

Biomimicry 2:
Thinking Like an Ephemeral Tepary Bean

Tepary beans—desert relatives of pintos, limas, and runner beans—are among the most drought-adapted annual legumes on earth.[14] Just 30 days after germinating, tepary plants begin to flower; they can produce a harvestable crop of dry beans in just 55 to 65 days under desert conditions. As such, these plants use the drought-avoidance strategies of many desert wildflowers by growing only during the brief summer monsoon season when soil moisture is available in North American deserts, and then producing a yield and dying before drought returns.[15] As we will discuss in chapter 7, they are sometimes classified as a drought-evading "ephemeral" rather than an annual crop plant, since their entire life cycle is so much shorter than the frost-free growing season.

What design elements allow tepary beans to withstand hot spells of 100 to 125°F (38–52°C) and to germinate, survive, and produce seed on as little as two hard rainfalls per season?

First, they put much of their energy into rapidly growing deep, water-absorbing roots rather than producing a lot of "excess" foliage. Their roots can reach depths of 8 feet in less than three months' time, and extend much farther belowground than the vines do aboveground. Next, their leaflets are much narrower and their flowers are smaller than those of other beans, so they have a smaller surface area exposed to the sun than do most bean plants. They also exhibit "self-shading," whereby their flowers and newer, tender foliage are often in the shadow of older, tougher leaflets. Internally, their proteins have higher **heat tolerance**, so they can bloom and set seed during hot spells that damage the leaves of pinto or lima beans and force their developing embryos to abort.[16] At midday, when air temperatures are above 100°F (38°C) and solar radiation can be brutal for other beans, tepary leaflets simply angle their surfaces away from the sun, but most of the rest of day their leaves behave like solar trackers—that is, they shift their angles to maximize photosynthetic gain as much as possible (see chapter 4).

Curiously, tepary plants require a mild drought or hot spell to trigger flowering and fruiting. When they have been grown in the past in wetter, cooler areas such as Minnesota, they seldom receive such a trigger, so they continue to produce as much green foliage as an alfalfa plant might in an irrigated pasture. When they are prompted to flower and fruit, they behave much like mature century plants, converting most of their vegetative biomass to seeds and pods, literally killing themselves while trying to reproduce.

Tepary beans have been grown by more than 30 native and immigrant cultures residing in the North American deserts of Mexico and the United States, which value them for their low water use, flavor, and nutritional value.

Eco-Mimicry: Thinking Like a Semi-Arid Prairie

Eco-mimicry is somewhat of a different dance from biomimicry or ethno-mimicry, for it requires protracted attention toward the way in which members of a plant and animal community work together to produce a variety of foods in a self-sustaining manner. Eco-mimicry allows us to imitate ecological relationships, habitats, and ecosystems—not merely the adaptations of a single species.

As noted earlier, my friend Wes Jackson is the internationally celebrated master of thinking like a prairie, and his attention to the pioneering work of Aldo Leopold, Paul Sears, and John Weaver helped guide him as he first formulated his vision. Of course, his three decades of weekly conversations with the poet/farmer/philosopher Wendell Berry certainly didn't hurt, either.

How to Think Like a Perennial Food Producer

In "thinking like a semi-arid prairie," Wes Jackson, Stan Cox, Tim Crews, and others have set goals for themselves that most agronomists and annual crop geneticists ignore, neglect, or fundamentally fail to consider when domesticating native perennial plants:

- The elimination of tillage;
- The sequestering of carbon in the soil;
- The fixing of atmospheric nitrogen in the soil;
- The enhancement of resilience through diversification; and
- The building of regenerative capacity without excessive inputs of fossil groundwater and fossil fuels.

These principles can actually be applied to redesigning agriculture for any region, not just for prairies.[17] They not only build upon "the perennial wisdom" of the prairie or other biotic communities, by emphasizing the hardiest edible natives in each, but they mimic the structure and function of those plant communities.

For my part, in the book co-edited by Wes called *Meeting the Expectations of the Land*,[18] I proposed a desert-derived design for drought-adapted polycultures. It intercropped cultivated varieties of columnar cacti, century plants, tree legumes, herbaceous perennials, and quick-maturing ephemerals that collectively mimicked the diversity of growth forms of plants found in the Sonoran Desert rather than relying on herbaceous perennials alone (see below).[19]

During a blazing-hot summer in 1979, I took a bus ride across the West to spend a few days at the Land Institute with Wes Jackson and his soon-to-be Sunshine Farm manager, Marty Bender. Despite being in the midst of a heat wave that scorched soybean, corn, and wheat fields nearby, the prairie outside of Salina, Kansas, remained resplendent with wildflowers and resilient in the face of so much heat stress. Over the course of the following decades, I've frequently returned to the Land to speak, and have remained friends with Wes and his daughter Laura ever since.

The Sahara's Ancient Food Forest: The Desert Oasis of Siwa

Of all the places I've documented food plant diversity on four continents, I will never cease to feel utterly enthralled by the food forests at the Siwa oasis in the midst of the Sahara near the Egyptian–Libyan border.[20] The Siwa oasis harbors more than a quarter million date palms, at least 25,000 olive trees, and enough tree fruits, vegetables, and grains to provide most of the sustenance required by 12,000 Berber and Arab-speaking residents. Drawing on the 200 springs and shallow wells in the Siwan Depression, farmers direct artesian waters through open (but often shaded) ditches to "waffle gardens" (see chapter 3). These gardens and adjacent date groves were traditionally flood-irrigated frequently enough to relieve the already alkaline soils of additional accumulation of mineral salts.

At least 33 crop species are still found in Siwa's gardens, groves, and fields; most of them are heat- and salt-tolerant perennials. These "food forest" perennials include almonds, bananas, carob, dates, figs, guavas, hibiscus, lemons, mangoes, olives, pomegranates, strawberries, taro, and many other tropical as well as Mediterranean food plants.[21] They are represented by many dozens of ancient heirloom varieties that are well adapted to the Saharan climate, in addition to 14 or so recent introductions that must be planted in cooler, shadier areas. Within each basin or "hod" in the waffle-like design, Siwan farmers intentionally place particular herbs, vines, bushes, or trees at measured spots along a gradient of salinity, from the middle of the basin to its slightly elevated edge. In addition to this horizontal zonation, the oasis farmers create a vertical zonation as well, with shade-loving vines and herbs situated immediately below the dense canopies of date palms and olives, where they sometimes use the taller plants as trellises and windbreaks. Of course, other crops are relatively shade-intolerant, and so many herbs and vegetables are situated in microclimates where they can be fully exposed to the sun.[22]

Sadly, recent tourist resort developments have been perforating more wells at Siwa, leading to more local competition for fresh water, and more ponds and standing puddles of brackish tailwaters.[23] Nevertheless, Siwa has outlived many other threats over its thousands of years of sustaining a rich, oasis-adapted agriculture, and my hope is that its Berber and Bedouin communities have the resolve and the resources to express their resilience once again.

Of course, desert oasis food forests are no longer restricted to the Old World, but over the last four centuries have been developed in the Americas as well. When Jesuit, Dominican, and Franciscan missionaries came from Europe to colonize Mexico, they initially brought desert-adapted Old World crops to the desert oases of Baja California. Of the 21 perennial food crops initially introduced to that desert peninsula before 1774, nearly all of them have persisted at sites where the missionaries mimicked the oasis structures they remembered from their travels in the Mediterranean and in the Canary Islands.[24] In essence, they emulated the classic model of oasis food forests that had developed in the Middle East and North Africa, then transplanted and refined that model to fit the hotter, drier climates they encountered in America's most arid regions.

Since those initial introductions from the Old World took hold in the Sonoran Desert, the California culture of Baja California has further diversified the food forests of desert oases. One rather remote Baja California oasis, San Javier, now harbors well over 40 varieties of 29 species of perennial fruit, nut, root, and herb crops, in addition to many annual vegetables and grains. Among all of the dozen or so oases historically colonized by European missionaries, more than 90 distinct species of perennial food crops from both

Rain gardens can benefit if you locate brush weirs upstream to slow runoff velocity and force the deposition of nutrient-rich organic matter. DRAWING BY BARBARA TERKANIAN AND PAUL MIROCHA.

the Old World and the New have been integrated into food forests that mimic the classic Saharan oasis agricultural designs.[25]

The take-home message for producing food under hotter, drier conditions is clear: To creatively emulate a time-tried model for oasis farming, you must pack as much diversity as possible into a vertical space, rather than trying to literally imitate what might be best suited to ancient Old World oases. The diversity brought in under the protective canopies of date palms and olives should be dynamic through space and time if it is to provide resilience in the time of rapid climate change.

Our friendship blossomed over conversations about the domestication of herbaceous perennial crops.[26] At the time, I was working as part of an Arizona team attempting to domesticate buffalo gourds, a wild relative of squashes, while they worked on domesticating eastern gamagrass, a wild relative of maize. But over the years, our conversations deepened. We debated how we could design resilient agricultural ecosystems on the models of prairies, cactus forests, and temperate rain forests. *Nature as measure and model* became our shared rallying cry.

Many students have grown so excited by the Land Institute's initial successes in domesticating wild perennial grains, legumes, and sunflowers that they literally can't see the prairie for its grasses. To me, Wes Jackson offers more than his finely honed expertise on perennial plant breeding (certainly a genius loci that deserves celebration); he is also an agro-ecologist who has contributed much to conversations steeped in reinventing food production systems in the face of climate change.

Whether the plants in his agro-ecosystem designs are fully domesticated or semi-cultivated, none will be grown in monoculture in the Land Institute's designs, nor will they include genetically narrowed annuals that require the incessant tillage, fertilization, and pampering that most industrial farming has demanded over the course of the last half century. Ideally, 12 to 15 different harvestable food species from several different plant families will be grown together in a single field to provide the highest yield stability in perennial polycultures suited to the conditions of the prairie biome—the landscape that stretches most of the way across the honey-colored heartland of America.[27]

Ethno-Mimicry:
Thinking Like a Desert Oasis

Modern permaculturists were not the first, nor will they be the last, to think of planting sets of temperature-sensitive plants as "guilds" in the shady canopies of fruit, pod, and nut trees. Traditional desert farmers have been doing this for centuries, from the Gobi in China to the Sahara in Africa all the way to the Sonoran in Baja California. Perhaps they initially mimicked the vertical structure of small trees, shrubs, and herbs that grew beneath palms and olive trees, finding species with edible fruits, nuts, or leaves to propagate beneath date palms.

Whatever the cultural origin of desert oasis agriculture, there is now an archipelago of island-like oases from Persia through Morocco that has harbored true food forests continuously for upward of 1,400 years. This is all the more impressive if you consider how easily salinity and alkalinity can accumulate and degrade soil quality around the artesian springs that water most desert oases. It may be that the vertical stacking of one kind of leafy canopy over another, and one kind of root system under another, reduces or processes the accumulation of minerals so that the associated food plants can remain productive.

Planning and Practice:

Planting Your Own Fredge to Stabilize Your Watershed and Enhance Your Foodshed

I'd like to give you an exercise to show how bio-, eco-, and ethno-mimetics can be applied to mitigate the effects of and allow adaptation to climate change. I'd like you to explore your own food-producing landscape for places particularly vulnerable to floods, soil erosion, or droughts and design a living fence, hedgerow, or vegetated margin that can better buffer your garden, field, orchard, or pasture from the consequences of extreme climatic events. You can use designs from nearby native plant communities or from neighboring cultural landscapes as a model for your plantings. First, though, I recommend that you walk around your home and look at topographic maps as well as aerial photos of it. Then ask yourself the following questions in order to plan your plantings:

- What micro-sites seem particularly vulnerable to winds, rains, floods, fire, or droughts on your property?
- How are they affected by pressures emanating from adjacent lands (for instance, the wounds from gully erosion back-cutting upstream)—and which of those pressures are within your control?
- What natural plant guilds or clusters of perennial vegetation on or near your site suggest or demonstrate a means to better buffer your landscape from these erosive or disruptive forces?

For those of you who have a garden, field, orchard, or pasture on a floodplain near a watercourse, here are the steps you can take to create a *fredge*, a living fencerow or hedge. If it is modeled on the Sonoran-style field edges, you can make it out of cottonwoods or willows of any species, with any sort of brush woven between the trunks. In landscapes or regions where cottonwoods and willows are less common, other kinds of riparian trees can be propagated by planting live stakes. Here's what you do:

1. Identify an eroded place on the floodplain alongside the edge of the streambed where you would like to protect or restore the remaining arable areas. Decide the length of your desired fencerow and determine whether it would be most effective as a linear or curvilinear streamside planting.

2. Between late December and mid-February, find a patch of dormant willow and cottonwoods (or other trees) nearby. Prune off three or four roughly 3-foot-long cuttings that average 2 to 5 inches in diameter, stripping them of their side branches and leaves. Ideally, aim for a ratio of 12 to 14 willows for every one to two cottonwoods that you prune and prepare as living stakes.

3. Dig a 2-foot-deep, 2-foot-wide trench 1 to 3 yards from the edge of the field (or other planted areas), and plant the cottonwood and willow stakes at 18- to 24-inch intervals. Fill the trench with moist sand to prop up the stakes.

4. Take the smaller pruned-off branches and weave 4- to 6-foot-long segments horizontally in between the upright (living) stakes until they create a sieve or webwork of materials 3 feet high from the ground.

5. Wait (and if necessary, hand-water) until late March, or at least until the live stakes root and leaf out as actively growing saplings.

These cottonwood stands were eco-mimicked by Sonoran farmers to maximize the benefits from their living fencerows or fredges.

6. After the first floods come, spread the deposited soil and flood-washed detritus out behind the fencerow, up against the existing field edge. Remove larger debris from both the expanded field area and from the fencerow itself.

7. On an annual basis, in mid- to late winter, reprune or coppice the cottonwood and willow saplings, and add new branches to the weir-like webwork among the saplings.

8. Every third year, thin the number of saplings as they grow in diameter, and trim the highest branches you can reach.

9. After every major flood, determine whether you should extend the existing fencerow upstream or downstream, or initiate another row 3 to 5 yards out toward the center of the watercourse from the last existing one.

With the planting of your own living fencerow, you have now joined the ranks of those who practice the ancient art of biomimicry worldwide. The further use of biomimicry, eco-mimicry, and ethno-mimicry will be demonstrated in the following chapters over and over again. Why? Because in the face of accelerated climate change, we need to diversify the designs we use for growing food, and for bringing it in from field to table,

Fig leaves offer a design for dispersing heat that biometrics designers can used for trellises that create a boundary layer effect.

with less of a carbon and water footprint than ever before. The various forms of biomimicry can help us diversify our options using designs that nature and various cultures have already field-tested on a particular scale, in a particular place. We must now experiment with them to see if they are adaptable to other places and scales.

Chapter Three

Will Harvest Rain and Organic Matter for Food

Catching Runoff as Conventionally Irrigated Agriculture Collapses

You don't miss your water
'Til your well runs dry.
      —WILLIAM BELL (Stax Records, 1961)

Warm-Up

Whenever I hear my Hopi friend Vernon Masayesva repeat to multicultural audiences his people's proverb, "Water is life," I try to fathom how much of the fresh, sweet water on this planet is already tied up in the lives of just one species, *our* species. Every second, 52 million gallons of water are used on earth to grow food and fiber for more than seven

The Turpan Water Museum in China, where visitors are reminded that harvested rainwater is the source of desert life.

billion other people who share the earth with us. Most of it is used for agricultural irrigation of food crops that we eat directly, for forage crops eaten by livestock, for agrichemical crops (such as the GMO maize that produces high-fructose corn syrup), or for flushing salts away from the alkali-saturated surfaces of our fields and orchards.

While it may take as little as 300 gallons of water to meet the daily dietary needs of a vegan or a wild forager, estimates of the water embedded in a meat-eater's diet range as high as 2,500 gallons a day. If we take 800 gallons per capita per day as the average amount of water needed to adequately feed all of our human kin currently dwelling on this planet,[1] we daily usurp as much as 5,620 billion gallons from other species to produce our own food and beverages.

When we add up the entire volume of water embedded in the agricultural products that we annually produce on this planet, 6.39 trillion cubic meters (about 5.6 million acre feet) of fresh water goes to feed and clothe us that might otherwise go to nourish and shelter other species, from desert pupfish to polar bears. This fresh water might otherwise flow into streams, rivers, lakes, and marshlands or go to support the wild vegetation in terrestrial wildlife habitats. A good portion of this enormous volume of "liquid currency" comes in the form of the surface water that has long been diverted from rivers and their reservoirs to irrigate crops and water livestock. Nevertheless, in less than a single century, farmers have extracted more fossil groundwater than all the water pumped from underground aquifers over the rest of human history. Strangely, the squandering of fossil groundwater has been done by pumping it with quantities of "cheap" fossil fuel so large that we may never see the likes of either of them again.

And if it's difficult to feed over seven billion mouths with water when it's relatively abundant and cheap, what will we do when accelerated climate change makes potable water scarce and rather expensive?

As Planet Water expert Sandra Postel has often reminded the many audiences she has spoken to over the last decade,

> All told, reaching the food production levels needed in 2025 could require up to 2,000 cubic kilometers [1.6 billion acre-feet] of additional irrigation water—a volume equivalent to the annual flow of 24 Nile Rivers or 110 Colorado Rivers . . . Supplying this much additional water will be difficult . . . The modern irrigation age—characterized by the engineering of whole river basins and mechanized control over vast quantities of water—is running out of steam. The best sites for dams are taken; numerous groundwater

Water and Food Justice:
Per Capita Water Use and "Embedded Water"

Several years ago, I was both humbled and surprised to learn that as late as the 1970s, desert-dwelling Native Americans such as the San Xavier O'odham of Arizona were using as little as 50 gallons of water a day to meet their basic needs, while residents in nearby Phoenix were consuming as much as 300 gallons a day, including their flood-irrigation of lawns and filling of swimming pools.[2] As native desert landscaping became commonplace in the more conservation-oriented neighborhoods of Tucson, per capita water used plummeted from 205 gallons per day in the 1970s to the range of 106 to 148 gallons three decades later.[3] In essence, newcomers to the desert are likely to consume four to six times the water that traditional desert dwellers have historically consumed. Excessive water use has therefore become an environmental and social justice issue.

But I was even more humbled when I learned that I had not taken into account—nor had most water analysts in the Desert Southwest—the water embedded in the production and transportation of food. In the United Kingdom, for example, 73 percent of a British citizen's annual water footprint is attributable to his or her foodprint.[4] Compared with the water each of us directly drinks each day—1 to 1.3 gallons—there is 3,250 times more water embedded in the food we eat! The farther fresh produce travels, the more times it is misted as it sits on trucks and in refrigerated displays, so the higher the water use of extra-local produce. According to the Institute of Grocery Distribution, there may be as much as 3 gallons of water embedded in a large (greenhouse) tomato, 10.5 gallons in a cup of instant coffee, 31.5 gallons in a glass of wine, 55 gallons in a glass of milk, and 630 gallons in a hamburger from a cow raised in an irrigated pasture or feedlot. While the amount of water and carbon embedded in any particular kind of food product varies dramatically depending upon whether it was grown with rainwater, pumped groundwater, or diverted river water—and in either a desert or temperate forest setting—all of us consume far more of our daily water budget through the food we eat (65 percent of the total water budget of our species) than through drinking, bathing, washing dishes, or growing houseplants!

reserves are already over-tapped. Reservoirs are filling with silt. Fertile soils are slowly being poisoned by salt. In short, our irrigation base is showing numerous signs of vulnerability just as we are about to become even more dependent on it.[5]

Of the 45,000 large dams built around the world to store fresh water for irrigating crops, many have become debilitated by accelerating rates of siltation and reduced snowpack diminishing the amount and duration of water flow into their reservoirs. The many consequences of climate change will stress and ultimately weaken these structures in other ways as well.[6] More recently, Postel has emphasized just why our irrigation base has become so vulnerable at the very moment that more humans than ever before are addicted to it:

The challenges loom large. It takes about 3,000 liters [793 gallons] of water to meet one person's daily dietary needs—about 1 liter per calorie. Satisfying this dietary water requirement for all—in the face of rising population and consumption, persistent poverty, and global climate change—will take a commitment well beyond what has materialized to date.[7]

Of course, some skeptics have attempted to dismiss Sandra Postel's projections as doomsday prophesies, just as they did with the late Marc Reisner's predictions in his classic *Cadillac Desert*, first published in 1986.[8] But when an illustrious team of water scientists returned to Reisner's predictions a quarter century after they were made,[9] they found that nearly all of them had come true, and that several trends in water scarcity, siltation in reservoirs, and declining irrigation water quality had become even more dramatic than Reisner had imagined.

In particular, Marc Reisner had originally suggested that the insidious buildup of salt in agricultural soils would become a significant factor in limiting food production in the Desert Southwest. The 25-year retrospective found that farming in the Desert Southwest has not only become highly vulnerable to salinization compared with other parts of the United States, but also generated farm revenue losses that have become 10 times higher than those in the East. The scientists confirmed that the damage caused by salinity has already resulted in over $2.5 billion of farm-gate losses per year in the western states that Reisner called the Cadillac Desert. This means a diminished public return on investment within large irrigation districts such as the Central Valley of California. There, state and federal subsidies for irrigated farmlands have amounted to as much as a $217-per-acre-per-year investment, even when farmers raised crops that were valued at only $290 per acre in some years.[10]

The salinization of farmlands in California and other arid states has likely been aggravated by the more frequent droughts that appear to have come with accelerated climate change. During such droughts, farmers often have no choice but to use poorer-quality water and flush out salts from their field soils far less frequently. Ironically, such salted-up soils then require more irrigation water to sustain any crop production than they would have needed otherwise. This vicious cycle has further stressed the already limited water resources of our country's most arid agricultural regions.[11] Not only is the Colorado River over-allocated, but in most years since 1950, its waters never reach the sea at its delta on the upper reaches of the Gulf of California. But the Colorado is not alone among rivers for

How Water Scarcity Will Restructure Agriculture Around the World

There are many factors associated with climate change that are contributing to widespread water scarcity, and not only in arid lands. In the eyes of environmental determinists like Jared Diamond, these factors threaten to precipitate the collapse of the artificial "hydraulic civilizations" that have developed in the American West and other arid regions, as Diamond has speculated that they did at Chaco Canyon and other places in the past.[12] While I assume that most foodscapes will not suffer a Diamond-esque collapse, I can imagine that they will be radically restructured under the following pressures:

- Hotter summer temperatures will possibly raise evaporation rates by 13 to 18 percent by the year 2050 compared with common rates of evaporation in 2000, generating more demand for water.[13]
- Carbon dioxide enrichment of the atmosphere will increase photosynthetic rates, which will also generate more (transpiration) demand for irrigation by crops.
- Managing crops over longer frost-free growing seasons will require two to four more irrigations per year than before.

- More sporadic and unpredictable rainfall regimes could mean that fewer farmers dry-farm their commodity crops, and instead rely on computerized center-pivot sprinklers, or drip or furrow irrigation to ensure that their crops receive a steady supply of moisture.
- Where careful estimates have been done (as in southern African watersheds), storm runoff into rivers is already decreasing, and will likely decrease further, by as much as 10 to 30 percent over our lifetimes.
- Should climate change trigger more frequent droughts or human population movements, it may place additional pressure on underground aquifers and surface water resources. During such crises, any water that can be easily or cheaply transferred from agriculture to urban populations may be shunted away from food production—and seldom, if ever, has such water ever been returned for irrigating farmlands. In this sense, usurping water from food production is never truly done in the name of water conservation; it is simply used to temporarily quench the thirst of burgeonimg urban and suburban communities.

generating saline soils and a dry delta; the Nile, Ganges, Huang He (or Yellow River), Amu Darya, and Syr Darya have borne the brunt of climate change and over-allocation. In fact, they are already so overtapped and stressed by droughts that little or none of their fresh waters reach the oceans anymore.

In the august pages of the Proceedings of the US National Academy of Sciences, Mark Howden and his colleagues have suggested that minor innovations to "tweak" agriculture toward greater sustainability will work only if forthcoming climatic changes are minor or moderate. But if climate change becomes as severe as some scenarios predict, they contend that merely "tweaking" conventional agriculture and global food distribution networks will have limited, if not negligible effectiveness. Instead, How-den and his allies suggest we should prepare ourselves for more radical restructuring of our food systems.[14] By paraphrasing and recombining the strategies for restructuring they have suggested, I have attempted to get at the efforts that can be made at any level of food production, from garden to field to farm to agrarian landscape:

- Shift from the extraction of groundwater and damming of large rivers into reservoirs to localized water harvesting and stream diversion.
- Shift to more low-water-requiring crop varieties and species (espe-cially perennials) that can tolerate drought, heat shock, and reduced winter chill periods.
- Vary the means to keep moisture retained in agricultural soils once it ar-rives there, reducing deep infiltration (subsoil drainage) and evaporation.
- Better manage surface water to reduce salinization and waterlogging of agricultural soils.
- Diversify farmers' portfolios of income-generating streams, by harvesting wild and managed products and receiving payment for ecological and social services provided by the land.
- Reduce the loss of harvestable food to pests, diseases, weeds, and waste.
- Rely more heavily on climate forecasting and monitoring of reservoir volumes, to cut back on the planting of annual crops when water is scarce and rains do not appear to be forthcoming, as means to reduce irrigation and conserve water.

In the next parable, we shall correct a common oversight made even by scientists as esteemed as Mark Howden: Harvesting rain and capturing run-off is not merely about capturing a sizable volume of water for irrigation, but also about simultaneously harvesting water, organics, nutrients, and seeds!

Parable

Three decades ago, an elderly farmer in the Sonoran Desert surprised and disoriented me by using a phrase I had never heard before. His name was Jerome Ascencio, and he had been involved for more than 50 years in tending and irrigating crops near the US–Mexican border, exchanging traditional farming knowledge with Anglo, Hispanic, and Native Ameri-can farmers and farmworkers. He himself had labored for many years as a farmworker in large pecan orchards and cotton fields fed by pumped groundwater and diverted river water. But when I knew him, Mr. Ascencio tended his own field, fed exclusively by storm runoff generated by summer monsoons and channeled onto his land from the ephemerally flowing streams that we call desert washes or arroyos.

Because Mr. Ascencio was entirely dependent on sporadic storms to provide the moisture he used to grow his food, he was unusually attentive to how the timing and intensity of rains had shifted over his lifetime. And so, when he conceded to me that he thought *the rain was dying*, I listened up:

"Rain, that's the main thing in the desert . . . ," Mr. Ascencio said to me in a voice that was hardly a whisper while he looked up at a cloudless sky. He was thin and tired, as were his clothes; he wore faded, bleached-out blue jeans that barely hung on his 20-inch waist, a twice-mended snap-button western shirt, old cowboy boots, and a straw hat that brought a long shadow down across his wrinkled face. He shook his head and sighed. "You can't just plant anything without the rains coming, without those washes running."

He then explained to me that the advent of the rainy season was running a month later than its usual onset. And so he had kept his vegetable, corn, and bean seeds in a wheelbarrow under a shady ramada rather than planting them, even though it was now late July. In short, terse sentences—for English was the last of three languages that he had learned over his lifetime—Jerome Ascencio tried to explain how something in the world had begun to go wrong for him:

"Now the rain is dying. Sometimes I feel so sad. I just stay here waiting . . . I don't do anything until those rains come . . ."[15]

And yet, rather than throwing in the towel and leaving his field fallow, Jerome Ascencio was intent on shifting his efforts to changing conditions, or as Sandra Postel puts it, "adapting to the new normal." When three storms came in early August, his field received enough sheetflow of stormwaters to make the soil moist enough for planting. He had carefully situated his half-acre field so that it captured the runoff and compostable plant matter from 30 acres of desert slopes just upstream from it. The nutrient-rich floodwash that was deposited by floodwaters entering his field was the same compost-like soil amendment that his neighbors in Mexico called *abono del río*. (This organic fertilizer and soil amendment will be further discussed in chapter 6.) Although he had never heard the phrase *water harvesting* at any point in his 70-some years of living, it was his livelihood, his love, and he was good at it.

What fascinated me most about Mr. Ascencio's responses to that drought was that he was so eager to shift and diversify his strategies to

A *de temporal* or *ak-chin* (arroyo mouth) field harvests runoff from adjacent canyon slopes in the Sonoran Desert.

ensure that his corn, beans, and vegetable crops received adequate moisture. First, he asked me to help him dig a small reservoir just upstream from his field. When it filled up on a single rain, then quickly drained as the waters soaked into the surrounding sand, he asked me to help him line the pond bottom with a sheet of plastic and bentonite clays. This small reservoir provided him with an opportunity to hand-water seedlings between summer rains, after the soil moisture had burned off from the field surface.

But Mr. Ascencio did not stop there. We brought truckloads of composted leaf litter in from where we found it deposited on the ground beneath mesquite trees on the long dirt road into his homestead. He incorporated this nitrogen-rich composted matter into his field soils just as he did the leaf litter that flowed in with the sheets of runoff. On occasion, he also added guano taken from a nearby cave, hoping to build up both nutrient levels and moisture-holding capacity.

It has always struck me that this desert elder—even in his late 70s—did not think of himself as a passive victim of drought or climate change, despite his sense of grief that the rains were dying. Instead, he responded with effective innovations that offered him more options to gather and hold water in and near his field, and to increase the capacity of his field soils to make the best use of that moisture right where his plants grew.

Is Water Harvesting *Only* for Small-Scale Food Production? Not at All!

Until I visited the Turpan Basin on the edges of the Gobi and Taklamakan deserts in western China, I erroneously believed that rainwater and runoff harvesting could never supply irrigation for food production on a scale of more than 200 to 2,000 acres per localized watershed. But in and around the ancient settlements of this basin, farmers across the centuries had dug and directed dozens of snowmelt-collecting underground conduits into 50,000 acres of vegetable, grain, and grape production without tapping into deep aquifers, rivers, or lakes for their water supply.

These subsurface *qanat*-like conduits or horizontal wells are called *karez* in local Uyghur and Hui Moslem dialects. The magnitude of runoff-generated crop production there is all the more amazing given the fact that most of the arid Turpan Basin is below sea level and receives less than an inch of rain in the average year, while its temperatures can soar as high as 108°F (42°C) during the summer. Evaporation rates in the Gobi are 1,000 times the volume of rainfall in the basins.

Two Uyghur boys in Turpan, China, show the opening of a *karez* horizontal well into a ditch that irrigates vineyards and vegetable plants with rainwater stored underground.

Breath holes for *karez* horizontal wells in the Turpan water-harvesting system near the Gobi Desert in western China.

A series of vertical wells or "breath holes" connect with horizontal *karez* wells leading to the vineyards and vegetable fields of Turpan, China.

A vertical well, with a bucket lift for water stored in Turpan, China.

Of course, *location* is everything when it comes to the harvesting of rainwater and snowmelt. The 20,000-foot-high ranges of the Tien Shan (and to a lesser extent, Flaming Mountain) receive considerable amounts of snow during the winter. As it melts, it cascades down more than 10,000 feet into the dry basin where loose alluvial sands have been naturally deposited over the millennia. The underground flows are particularly strong within arroyos or wadis that drain the flanks of Flaming Mountain, 5 to 8 miles from the city of Turpan. By first excavating vertical shafts that then serve as "breathing holes" every 10 to 25 yards, then removing sand, gravel, and rocks from tunnels carved out 20 to 60 feet below the surface, long "horizontal wells" are developed from the sandy alluvial fans below the mountains to just above their fields, where they open into irrigation ditches that run through hundreds of fields and gardens.

In this way, Uyghur and Hui farmers and gardeners grow food year-round by tapping into significant flows of snowmelt that have infiltrated just below the soil surface. Some of these farmers also themselves rest below ground at midday in subterranean "patios" that they have built along wider stretches of the *karez* tunnels, where summer temperatures may be 30°F (17°C) cooler than those out in the desert sun.

Some of their horizontal wells run for 5 to 8 miles above their planted areas and have been in use for as much as 1,500 years, since the *karez* system (or more broadly, the *qanat* tradition) was introduced from Persia and the Arabian Peninsula. A few of the newer *karez* wells run only a mile or two and use small electrical or gas pumps to lift water to the surface, but most of the water delivery remains gravity-fed and free of fossil fuel or fossil groundwater use.

In addition to 50 outstanding varieties of heat-tolerant table grapes, the Uyghur and Hui farmers in the Turpan Basin grow an astonishing diversity of vegetables, including eggplants, daikon radishes, squashes, yams, yard-long beans, watermelons, bitter melons, Persian melons, peppers, turnips, rhubarb, corn, sorghum, mint, garlic, onions, chives, leeks, and numerous

Lush flood-irrigated "waffle gardens" growing maize, millet, and yard-long beans with collected rainwater flows in Turpan, China.

The Flaming Mountains of the Gobi Desert provide sporadic runoff from their slopes to supply the *karez* water-harvesting system of Turpan, China.

greens for salads or for stir-frying. They also grow fruits such as mulberries, goji berries, apricots, plums, and table grapes. Their grapes are renowned throughout Asia, and are trained to grow up vertical poles to horizontal trellises 14 to 20 feet above the ground; these grape arbors provide shade to many of the walkways in the city of Turpan, where a quarter million residents obtain the bulk of their produce from the fields and gardens fed by the *karez* irrigation system.

If 250,000 desert dwellers can be primarily nourished by the harvests from 50,000 acres of intensively cultivated vegetables, grains, and fruits irrigated by capturing snowmelt and rainwater in subterranean tunnels, what else is possible in rainfall-limited areas of the earth?

The Lost City of Jiaohe near Turpan, China, flourished for centuries on its natural capital of harvested rainwater.

Principles and Premises

If the continued practice of conventional irrigation agriculture has become tenuous because of the water scarcity generated by climate change and the overdraft of aquifers, won't the potential for water harvesting also be constrained as shifting weather patterns diminish or alter rainfall patterns? While it is true that droughts can obviously affect the volume of rainwater that can be harvested in any place, it is also true that we hardly utilize even half of the collectible storm runoff available in most regions, and far less than that is conserved and targeted for food production. Even in the driest reaches of sub-Saharan Africa, where the extent of rain-fed traditional agriculture has not declined as precipitously as in other regions, an estimated 45 percent of storm runoff available in or near farmscapes is not currently harvested by farmers to sustain the soil moisture in their fields and orchards.[16]

In short, the gross amount of runoff that might be available to support food production will inevitably decrease with climate change, but regardless of that, farmers currently net only a minuscule portion of all the stormwater runoff that could be easily harvested to augment the growth of their food crops. And since the costs of pumping groundwater and building or maintaining dams are increasing exponentially, investments in water harvesting are becoming ever more cost-effective.

There are many water-harvesting traditions still intact around the world, in addition to recent modern technological innovations that advance the contemporary practice of this sustainable art form. Yes, an *art form*. Traditional water harvesters have combined their sense of aesthetics with their interest in utility to create remarkably elegant landscape designs to capture and deliver water to food plants. Their water-harvesting systems are often as eye pleasing as they are efficient.

Given that the premise of a solar-charged, rainwater-nourished food production system seems to be a viable one for reducing the ecological footprint of our food system, what principles associated with sustainable water-harvesting systems might help us adapt to climate change? Here, for starters, are a few:

1. The most sustainable strategies for harvesting rainwater will also serve to harvest nutrients (and effective soil microbes) in service to food production. In other words, capturing storm runoff from wildland

ecosystems not only provisions water but also deposits nutrient-rich compost and microbial cultures on our fields, gardens, and orchards. Capturing rooftop runoff and storing it in a tank or cistern does not.

2. The ultimate place to store water is in the soil immediately surrounding the food plants you have chosen to cultivate or harvest from the wild. That said, it may nevertheless be desirable to store water in multiple ways and in various places that provide backup water supplies between rainfall events. Funneling runoff directly into swales or waffle gardens of vegetables, fruit trees, or vine crops is highly effective, but providing them with supplemental irrigation from water stored in clay pottery **ollas**, as well as in cisterns, water-harvesting tanks, and lined reservoirs, should not be ruled out. (See chapter 4.)

3. The volume of water to be captured for food production is dependent upon:

   - The projected seasonal (or annual) rainfall volume and intensity;
   - The size of the "catchment area" or watershed above the field or garden;
   - The **runoff coefficient** of each surface in the catchment area (the percent of total rainfall shed from or infiltrating down into soil *above* your planted area);
   - The mix of catchment surfaces and the area of each in the watershed above your planted area; and
   - Any transmission, infiltration, and delivery losses that occur while transferring the water harvested from your catchment area to your planted area (leakages, overflows during intense storms, et cetera).

4. The drier the climate, the higher the ratio of catchment area to planted area will need to be. Traditionally in climates that averaged roughly 8 to 10 inches of rainfall per year, Hispanic, O'odham, and Hopi Indian floodwater fields drew upon watersheds or sloping catchment areas above them that were 12 to 60 times larger than the planted areas.[17]

5. With climate change, it may be prudent to reduce your estimates of water available to you from a catchment by 20 percent of the running average precipitation for your locale, not only because rainfall may decrease, but because transmission and delivery losses may increase with catastrophic weather events.

6. While it is possible to calculate the consumptive water demand for single crop species planted over large acreages, most water use estimates cannot easily be translated to intercropped species in polycultures or

A Rain-Fed Waffle Garden for Native Medicinal and Culinary Herbs

Rainwater-harvesting storage cisterns hold rooftop runoff until it is needed in nearby plantings.

When I suggested to landscape designer Caleb Weaver that I wanted his help in designing a rain garden of native herbs for my wife, Laurie, for our anniversary, he began to walk around our property looking for the places where rainwater and runoff naturally collected in the greatest quantity. Ironically, it was at a spot on the edge of our ridge that I had overlooked, for it seemed so overgrown and rocky. And yet, when I saw what Caleb was proposing to do, I realized that it would not only receive runoff spilling over from our rooftop rainwater collection cisterns, but also capture sheet flows and sediments trickling down a small watercourse from our courtyard and pollinator gardens.

Caleb crunched the numbers on the catchment-size-to-planted-area ratio, and found that we could deliver roughly nine times the amount of rainfall and runoff that would normally fall within the planted areas itself! And yet, knowing that there would still be three- to five-week periods where no rain fell, I selected drought-hardy native perennials that had long served

Brush weirs direct and slow runoff that flows into a rain garden. PHOTO BY CALEB WEAVER.

A rain garden designed as a "waffle garden" slows the flow of water and nutrients to grow perennial edible crops and medicinal herbs. DRAWING BY PAUL MIROCHA.

residents of our region as medicines, spices, and teas—Mexican oregano, oreganillo, Kota tea, desert lavender, skullcap, giant hyssop, chiltepin peppers, pipevine, and desert hopbush. These perennials quickly extend their roots deep enough to survive droughts of this length without any supplemental irrigation. We had just enough slope to lay out 12 "hods" of a waffle garden, line them with rocks, and mulch them. We planted each native herb with biochar, effective microbes (including mycorhizzae), and fish emulsion placed within their root zones.

On the slope above the waffle gardens, we constructed four brush weirs to slow down the runoff and catch nutrient-rich sediments drifting out from beneath the mesquite trees located upstream. We hand-watered the native perennials in the waffle gardens just three weeks before the summer rains came. Within another six weeks they had fully rooted out and doubled in size. The aromas from these plants after a summer rain are enough to make me weak at the knees. What's more, they remind us of a simple fact: *The desert can cure.*

permaculture designs. Furthermore, as evaporation rates increase as much as 10 to 15 percent over the next four decades, higher transpiration for leaf surfaces will inevitably lead to higher water demand per food crop species or variety, *unless* multi-tiered nurse plant guilds buffer understory species from the higher evaporative pull.

Water harvesting can be accomplished at any scale, but designs for smaller planted areas (anywhere from 50 square feet to half an acre) are often called **rain gardens** or infiltration basins, while fields and orchards of half an acre to 40 acres are often called floodwater farms, *de temporal* or *ak-chin* agriculture.[18] For the purposes of helping you initiate or refine your own practice of water harvesting in the landscape where you live, let us co-design a rain garden that fits your particular setting just as Caleb Weaver and I did in June 2012 in Patagonia, Arizona (see sidebar on page 70). You can now use new inexpensive apps to help you with the calculations and runoff coefficients required for this next exercise: Rain Harvest for iPhone, iPad, and iPod Touch, or Rain Harvest Calculator for Androids.

Planning and Practice:

Rain Gardens

A rain garden is a planted depression or infiltration basin designed to absorb most or all of the runoff generated from rooftops, hardscapes, or natural habitats on a particular property.[19] Downstream from a catchment area, an absorption area is planted to native vegetation, or to (usually perennial) edible or medicinal plant landscaping to soak up rainwater and storm runoff that might otherwise cause flooding, pollution, or erosion downstream.[20] Although the term was coined in the 1980s in the state of Maryland, similar gardens have been designed in arid regions since time immemorial, and include Old World eras and hods, which use floodwaters to nourish their garden plantings, as well as the waffle gardens of the Zuni Indians of New Mexico and traditional oasis farmers of Sonora and Baja California.[21]

There are several essential steps that must be taken before you can effectively catch rainwater for food production in waffle gardens, eras, and hods:

1. *Scope through on-site observations.* When it rains at your site, set your lawn chair out and watch. Watch where the stormwater flows with

different intensities of rain. Sketch the drainage patterns in your landscape. Allow your eyes to follow the water as it dances over different surfaces. See how it rushes around rocks to scour exposed earth, but moves more gently through grass and herbs.

2. *Record your observations.* After the rains, take note of where water has puddled, where newly sprouted seedlings volunteer, and what types of plants grow there. Pinpoint where water infiltrates into the ground. Chart the spots where it speeds on hard, impervious surfaces like concrete and asphalt, gathering in volume and eventually causing rill erosion.

3. *Assess your options.* Create a rain garden only where it makes sense on your property. Look for a large, relatively flat area on your site where water naturally runs. When it's not raining, look for clues that rain collects in this area: moist soil in the lowest places, pockets of debris, a heavy layer of mulch, or finer soil grains than on the majority of your site.

4. *Assess the soil types on your land.* Ensure that you have a site with enough drainage so that plants inundated by water won't be starved of oxygen, but not so much sand that all the water drains away.

A spillway or swale directs stormwater runoff downstream toward an orchard in southern Arizona.

A mesquite log is the final "speed bump" slowing rainwater runoff before it enters this space, which is being prepared to become a rain garden. PHOTO BY CALEB WEAVER.

5. *Refine the plan.* Create a basic structure for your rainwater-harvesting plan. Have your drainage ways follow the natural path of the water; don't make extra work for yourself or disturb the soil unnecessarily.

- Use berms and swales to direct the water to where you want it to go.
- Use brush weirs and **check dams** to slow water flows before they reach the planted area. (See chapter 8.)
- Dig and shape basins to capture the water so that it can infiltrate into the root zones of the plants you sow or transplant.

Boomerang-shaped micro-catchments work well for multiplying the rainwater and runoff that's available to fruit trees. DRAW-ING BY PAUL MIROCHA.

6. *Create added resilience in the face of climate change.* Plan for extremes! Storms may become less frequent in some places, and yet those storms that do arrive will likely increase in severity. Create water control structures for your rain garden that will absorb or divert all the water arriving in extreme weather events, or they will blow out your berms and do damage to plantings.

7. *Make sure there's an overflow outlet* through which you can divert excess runoff. Have a backup watering system in case of extended drought, especially if it should occur while you are establishing plants.

8. *Take time to crunch the numbers.* Assess the runoff coefficients for the surfaces in your catchment areas (see table 3-2), and determine an ideal ratio of catchment area to planted area to better predict available water yield to meet the consumptive water use of your desired plants. Again, the Rain Harvest calculator apps can help you here.

- Bring the data together to calculate your land's water yield: Multiply the total catchment area times the runoff coefficient for each surface times the annual average rainfall estimate for the catchment area above where you want to plant. For large catchment areas with porous soils along the drainage ways, calculate transmission losses as well.
- Measure your total catchment area, then measure the sub-area of each different surface type (bare clay soil, sidewalk, rooftop, asphalt or bitumen driveway or street, et cetera) and multiply

TABLE 3-1 Runoff Coefficients from Various Manmade and Natural Surfaces

SURFACE CLASS	SURFACE SUB-CLASS	RUNOFF COEFFICIENT
ROADWAYS	Asphalt (Bitumen) pavement	0.95
	Concrete pavement	0.80–0.95
	Brick paving	0.70–0.85
	Gravel or hard-packed clay	0.20–0.75
WALKWAYS	Brick paving stones with hard-packed dirt	0.10–.070
	Pavers with turf blocks in between	0.15–0.60
	Crushed aggregate	0.20
LAWNS	Turfgrass on flats, sandy soil	0.05–0.25
	Turfgrass on flats, clay soil	0.15–0.35
	Turfgrass on gradual (3–10%) slope	0.20–0.40
	Turfgrass on steep (10–45%) slope	0.20–0.45
NATURAL VEGETATION	Flats with dense vegetative cover	0.10–0.20
	Gradually sloping or rolling swales with modest vegetative cover	0.20–0.30
	Hillsides with patchy vegetative cover	0.25–0.40
	Steep slopes of mountainous ridges or canyons, little vegetative cover	0.30–0.50
ROOFS	Pitched metal roofs	0.95
	Pitched concrete or asphalt (Bitumen) roofs	0.90
	Built-up tar and gravel roofs	0.80–0.85

Source: Adapted from http://green.harvard.edu/sites/all/themes/green-ofs/theresource/leed-submit/nc/documents /Wyss_SSc6_1.pdf

each sub-area by its runoff coefficient and by average annual rainfall (in feet) reduced by 20 percent to factor in potential climate change. Sum up all sub-areas' square feet and multiply by their runoff coefficients and projected rainfall amounts to get a projected annual yield in cubic feet or acre-feet.

- Use reference books to determine the consumptive water use per year per acre for the crop or crops you wish to plant, then divide that by the projected water yield of your catchment area to determine the size of planted area you can construct and adequately water as your rain garden.

9. *Shape your catchment area and watercourses for runoff delivery to your planted area.* Collect tools and materials including a **bunyip** (a spirit level assembled from a transparent tube and two measuring sticks); flags or rocks for cairns; a mattock (a pick-ax with one end

TABLE 3-2 Consumptive Water Use Estimates for an Acre of Continuous Cover by the Foliage of Various Food and Fiber Crops

CROP	CROP WATER NEEDS (mm/total growing period)
Alfalfa	800–1,600
Banana	1,200–2,200
Barley/oats/wheat	450–650
Bean	300–500
Cabbage	350–500
Citrus	900–1,200
Cotton	700–1,300
Maize	500–800
Melon	400–600
Onion	350–550
Peanut	500–700
Pea	350–500
Pepper	600–900
Potato	500–700
Rice (paddy)	450–700
Sorghum/millet	450–650
Soybean	450–700
Sugar beet	550–750
Sugarcane	1,500–2,500
Sunflower	600–1,000
Tomato	400–800

Each sub-basin of a waffle garden should be designed to hold as much water as possible for deep soil penetration, but allow for overflow escape to the next sub-basin during heavy storms without eroding away its edges.

flat); a shovel; a digging bar; a tamper; loppers and pruners; locally available sandy loam; cobble-sized rocks; mulch; and compost.

10. *Remove rocks or prune any trees or shrubs that are halting, diverting, or disrupting flows to the planted area.* Shape wide, shallow drainages from the upstream catchment area to the planted area. If the slope is steep or predicted rainfall intensity high, place brush weirs or single rock alignments across the drainage way to reduce the velocity of water flows into the planted area, and to deposit silt over a broader area.

11. *Shape the planted area into the sub-basins of waffle gardens that are called hods or eras,* constructing drainages and introducing the plants: Shape the planted area into a series of 3- to 5-foot-wide sunken sub-basins with earthen walls on their perimeters 6 inches above the basin floor, and lined with 6- to 9-inch-diameter cobbles.

12. *Add compost, biochar, mycorrhizae, or effective microbes* to the planted area in each era or waffle, and shape the basin to hold water.

13. *Plant seeds or make holes for transplanted perennials* in each waffle basin, and water until rains begin. Make food from rain!

While rain gardens seem to be well suited to smaller watersheds in urban as well as more rural landscapes, the same concepts can be applied to truly agrarian and wilderness landscapes at all scales. I once served as translator for the National Park Service when it contacted O'odhan farmer and elder Delores Lewis to plan a series of brush weirs to manage runoff and halt erosion in a federally recognized wilderness area in Organpipe Cactus National Monument, not far from the Arizonan–Mexican border. He basically designed a plan to place brush weirs and other water diversion structures so that they could deposit soils rather than erode them in the aftermath of torrential desert rains. All this could be done by hand in a wilderness area where bulldozers, backhoes, metal and wire **gabions**, and other technologies were no longer allowed!

My point is simple: Depending on the size and condition of your property and its surrounding watershed, imagine how water-harvesting designs—from one-tenth-acre rain gardens to 50-acre floodwater fields—can heal wounded places, and nourish them back to a condition where they are capable of producing food from both wild and cultivated plants.

Chapter Four

Bringing Water Home to the Root Zone

Getting More Efficient at Irrigation Delivery

Warm-Up

Whenever it rains at my home on the edge of the desert, I hear it beat down on our metal roof, then resonate and rattle through a "rain chain" that drains the downspout from our gutters. It is music to my ears. As runoff from our rooftop cascades down a chain of cups and enters one of our cisterns, I am glad for our capacity to hold water so that we may deliver it to our vegetables days after the rain clouds have vanished, when thirsty roots are ready for another drink. But how, I wonder, can I best deliver that precious water to my chile peppers and tomatoes so that they get just the size of drink they need, and without wasting much of that hard-won water along the way? By sprinkler? By handheld hose? By flooding furrows, swales, or micro-catchments? By drip irrigation line? Or by buried pitcher—that is, a water-filled olla sunk into the ground next to their roots?

Once a water supply has been dedicated for food production— whether it is from rain, stormwater ponds, cisterns, wells, rivers, or lakes—it is amazing how many factors impede our ability to rapidly and fully convey that water to the root zones of food crops. As a desert farmer once joked to me, if fluids were always that easily delivered right to where they were targeted, no one would ever have to mop up the floor by a toilet in a bathroom! But water flows where it is not necessarily intended to go, and the phrase *transmission losses* has become a euphemism for the profligate waste of an already scarce resource.

If irrigation water is already becoming an ever-scarcer commodity with climate change, it is high time that food producers begin to redesign

Recycling graywater for watering crops can simultaneously help with its cleansing treatment and produce food, as is being done here near the Dead Sea in Palestine.

and fine-tune how they provide moisture to the roots of the food plants they wish to grow. We may need to realign gutters, pipes, sprinklers, spillways, or swales to reduce losses, perhaps even shortening the conveyance distance between our crops and their water sources. We will always need to periodically curtail leaks from pipes and canals, and unclog the screens on top and the nozzles below cisterns. We should mulch and cluster our crops to stem evaporation losses from bare soil. In essence, to become better food producers in the face of climate change, we must learn how to become better "water and food system plumbers," redesigning our water delivery strategies to best meet the needs of our food plants. Much of the work we do as plumbers will not be very glamorous or visionary, but it is as important as any other work we undertake in adapting our food systems to more austere conditions.

As Gary Pitzer of the Water Education Foundation once put it, "There are two constants regarding agricultural water use: 1) growers will continue to come up with ever more efficient and innovative ways to use water, and 2) they will always be pressed to do more."[1]

There is a peculiar paradox that surrounds our efforts to achieve greater food security under ever-drier conditions. As farmers have been forced to become more efficient in delivering water to the crops that feed us, urban dwellers have not been pressed to conserve water to the same extent. In the arid West of the United States, urban users increased their per capita consumption of water by 24 percent over the last quarter of the 20th century, while farmers and ranchers reduced their water use by 20 percent. In fact, farmers in the western US states have dramatically curbed their water withdrawals for irrigating crops, reducing both the acres irrigated and the volume of water used per acre. Much of the water savings occurred as farmers abandoned the practice of conventional furrow irrigation to deliver water to their crops (with a 45 to 65 percent efficiency in getting water to the roots) and instead embraced center pivot sprinklers (with a 75 to 85 percent efficiency) or drip irrigation (with an 85 to 95 percent efficiency).[2] Although these water delivery technologies were expensive to implement, reducing water use and pumping costs by a fourth to a third was certainly in a farmer's best economic interests over the long haul, even without the demon called climate change rearing its ugly head.

But while the national average of per capita water use is about 40 gallons daily, the average per capita use in the major cities of the Desert Southwest is three times as high, with some residents of the Sun Belt still using over 300 gallons a day. Throughout the 19 states in the western US, urban per capita water use increased from an average of 129 gallons per day in 1960 to 160 gallons a day in 1990.[3] Per capita water use in Phoenix, for example, increased from 246 gallons in 1985, to 253 gallons in 1990, to 259 in 2000. Only after Arizona's economy showed the first signs of its disastrous economic downturn with the onset of the mortgage crisis did per capita water use begin to decline, dropping to 216 gallons in 2008.[4]

But that is only a 13 percent decline in the average urban dweller's per capita water use after a quarter century of public agencies spending millions on urban water conservation education.[5] A 13 percent reduction over decades seems paltry, when Metro Phoenix is likely to lose 15 to 20 percent of its main supply of water from the Colorado River due to the consequences of climate change.[6] Because similar losses will most likely occur in all cities across the arid West, it is not surprising that we hear more desperate talk of engineering extremely expensive interbasin transfers to keep our cities alive in the future. The last plan proposed urges the federal government construct a 600-mile pipeline from Leavenworth, Kansas, to the Front Range region surrounding Denver, requiring that irrigation water be pumped uphill from the Missouri River into the Colorado River Basin.[7] While such costly delivery systems may move irrigation water closer to where populations are currently living, they in no way guarantee that this water will be delivered all the way to food crops or kitchen faucets with any efficiency.

However, when compared with the 30 percent gains in water delivery efficiency made by farmers in Israel in less than three decades, US farmers and urban dwellers with lawn sprinklers have yet to increase their water application efficiency all that much. In the early1960s, Arizona farmers used about 25 acre-inches per irrigation application per acre, but by the late 1970s they had reduced their irrigation applications to 20.5 inches, a decrease of almost 20 percent.[8] From 1980 to 1985, agricultural water use declined another 5 percent, but it dropped only another 2 percent by 1990 as easy gains simply based on shifting irrigation technologies became harder to come by.[9] Over the course of four decades, US farmers failed to hit the water conservation targets that Israeli and other Middle Eastern farmers achieved in far less time.

As climate change advances, simply continuing the shift from open furrow irrigation to drip irrigation will no longer be enough to keep farmers out of "hot water." Fortunately, there are a great number of alternative irrigation systems—traditional and innovative—suited to hotter, drier climates, and already being successfully used by farmers around the world.[10] One of the traditional strategies for efficient water delivery to food crops—buried clay pot irrigation—has well over 2,000 years of successful use in various deserts of the world, and is now undergoing a revival in the US Southwest.

Parable

In the summer of 1974, I learned of a remarkable achievement by a Mexican farmer in the Chihuahuan Desert from a legendary rural sociologist, Carl Kraenzel, who was then living in El Paso, Texas. Dr. Kraenzel had authored the classic book *The Great Plains in Transition* decades before, having devoted much of his career to delving into the effects of drought, water scarcity, and soil management during the Dust Bowl era. As a result of this research, he was extremely attentive to how farmers can either harm or heal a working landscape by their actions. When Dr. Kraenzel learned that I had spent time in the deserts of northern Mexico, he told me how much he had been impressed by the efforts of one particular Mexican farmer who had enhanced the food production value of a desert landscape just south of the border.

"After one of my lectures about the Dust Bowl, a young woman from Chihuahua came up to the lectern and asked if she could tell me about her father," Dr. Kraenzel recalled. "She noted that she saw no inevitability to land degradation at the hand of man, because her father had shown her ways to increase the productivity of the desert where she had grown up by very simple means. In fact, she gratefully noted, her father had made enough income from farming melons in the desert to put her and her two sisters through college. He did so without using conventional irrigation."

Dr. Kraenzel then described to me his visit with his student to her father's farm outside of Juarez, Chihuahua, on the remote edges of an enormous mass of sand known as the Samalyuca Dunes. There, in what Kraenzel referred to as the *yonland*, her father grew thousands of pounds of watermelons with a minimum of water at his disposal. To do so, he employed little more than inexpensive, unglazed clay vessels known as ollas, a bag of home-saved seeds, and some plastic water containers that he could fill at a nearby well and carry by mule or pickup truck to where the ollas were buried in the sand up to their necks. From the seeds he planted in the sand surrounding the ollas, he harvested a salable crop of melons each year, despite the fact that the area received less than 9 inches of rain in an average year, and often suffered summer temperatures as high as 110°F (43°C). I later learned from desert restoration ecologist David Bainbridge that melon yields can reach 25 metric tons per hectare using just 2 cubic centimeters of "olla irrigation" per hectare. That translates to nearly 11 tons of melons per acre on less than an acre-inch of water delivered through the pores of clay pots!

"This kind of innovative approach to water conservation accomplished by peasant farmers," Dr. Kraenzel confided in me, "should humble those of us who call ourselves scientists . . . It's the kind of traditional knowledge that we should take keen interest in, and document to the extent that is possible."

Within a few years of meeting Dr. Kraenzel, I began using and promoting Mexican ollas to provide moisture to crops as diverse as Seminole pumpkins, yellow-meated watermelons, and jack beans. With my New Mexican colleague Sam Hitt, I published one of the first articles in English calling for a revival of buried pottery pitchers in dry lands.[11] From responses to that article, I soon learned that vegetable crop irrigation by buried clay pitchers had been described by Fan Sheng-chiu in China more than 2,000 years ago, and practiced from the Gobi and Taklamakan deserts in Asia, to the Sahara in North Africa. No one knows when this practice came to the Americas, or even whether it could have been independently reinvented in

The use of *olla* irrigation using fire clay water pitchers is regaining momentum in Arizona and New Mexico, as in other parts of the world. DRAWING BY PAUL MIROCHA.

the deserts of Mexico. Today, however, it is practiced by traditional desert farmers in China, Pakistan, India, Burkino Faso, Zambia, Zimbabwe, Brazil, and Mexico, as well as in Arizona, California, and New Mexico.[12]

Principles and Premises

For the tremendous physical, financial, and even political efforts that farmers undertake to secure reliable irrigation supplies in order to produce food in a time of water scarcity, it is amazing how sporadic on-farm

Olla clay pottery irrigation is an ancient technique, first practiced in China, that has spread through ethno-mimicry throughout the world.

follow-through can be that might ensure more efficient water delivery to their crops. Once larger-scale farmers have taken the time and invested the money in laser-leveling their fields, installing center pivot or automated drip technologies, they may be either so broke or so exhausted that they do not have the resources to fine-tune their systems so that they continue to conserve more and more water. That is why ecologist David Bainbridge suggests that we must look *beyond* drip irrigation to various other means of **micro-irrigation** that do not demand such high maintenance and peri-odic technological replacement costs.[13] These micro-irrigation strategies include not only the buried pitcher irrigation described above, but also **wick irrigation**, porous capsule irrigation, deep pipe irrigation, perforated

Wick irrigation from small reservoirs of freshwater kept in covered buckets or containers is one of the most water-efficient means of micro-irrigation. DRAWING BY PAUL MIROCHA.

drainpipe irrigation, and porous hose irrigation, to name a few.[14] They can and should be complemented by mulching, micro-leveling, propagating nurse plant guilds or tree shelters, building micro-catchments or wicking beds, and reducing transpiration losses from weeds.

When thinking about how to effectively deliver water from a well, reservoir, or storage tank to the roots of a plant, keep the following rules of thumb in mind:

- The farther away the plants are from the water source, the greater the risk of water loss from leakage, breakage, evaporation, and other disruptions of the conveyance system.
- The greater the lift required to move water vertically from a well, underground storage area, or stream below the elevation level of the planted area, the greater the pumping costs per foot of lift, so that fossil fuel costs and carbon footprints may increase exponentially.
- The broader and less directed the water flow is toward the plant root zones, the higher the risk that water being targeting for the crops is missing its mark and supporting weeds, diseases, pests, or salinity buildup rather than food production.
- The greater the surface area of bare soils found in your field, garden, or orchard, the higher the frequency of evaporation losses.
- The hotter the temperature and higher the sun is in the sky when you irrigate, the greater the evaporation losses. The hotter the climate, the faster the deterioration rates and breakage of plastic and PVC piping.
- The greater the use of plastic, PVC, and other petrochemical-based materials as vessels for transmitting water, the greater the risks of higher carbon footprints.
- The less sorted your crops are into different **hydro-zones** based on their physiological needs for moisture, the greater the potential for underwatering or overwatering some of them.
- The poorer the timing of planting and harvesting is in relation to the timing of rains, the longer the duration of supplemental irrigation.
- The greater the salinity in a field soil, the more water will be required to flush the salts from the soil surface.

Fortunately, many of these risks can be simultaneously dealt with through the more holistic planning and design strategies characteristic of permaculture and other agro-ecological approaches. For instance, if swales are properly designed, nurse trees and their understory herbs can be clustered by hydro-zone, and pruned branches from the nurse plant species can be used to mulch any otherwise bare soil. Selecting desert-adapted, short-cycle vegetables, grains, and legumes for a planting that is timed to coincide with the advent of summer rains will reduce the frequency of water conveyance needed, as well as reduce losses from evaporation, transpiration, and percolation. Shallowly burying any piping or hoses—or running them primarily under the shade of perennial plants—reduces solar radiation damage, rates of deterioration, and replacement costs.

Conserving water through reducing conveyance losses may seem like a left-brain exercise relative to the creative design of water-harvesting systems that the right side of the brain loves to do, but it is just as important and provides its own opportunities for creative problem solving as well.

✿

Planning and Practice

Let's look in detail at buried pitcher irrigation, a water delivery strategy that can register as high as 95 percent efficiencies in conveying moisture to the root zones of selected crops. The steps in setting up such as a strategy are as follows:

1. Obtain (hopefully from locally produced sources) unglazed pottery or terra-cotta vessels that have been fired below 1,800°F (1,000°C) and are capable of holding anywhere from 0.75 gallon to 2 gallons of water. Squat, pumpkin-shaped pots with thick flared rims are ideal, but many shapes can be used.
2. Select a site—preferably with sandy rather than dense clay soil—and dig a hole twice as deep and three times as wide as the pottery vessel you wish to bury.
3. Mix compost, sand, biochar, and aged manure in the lower third of the hole. Then place a flat piece of clay, terra-cotta, or ceramic (a saucer, drainage tray, or broken piece of another vessel) at the base of the hole, immediately beneath where you wish to place the pot.
4. Position the pot on the saucer or tray to impede water loss from the bottom, with the height of the pot allowing its neck and top rim to emerge an inch or two above the land surface.
5. Fill the pot full of fresh water—add no fertilizer or manure to the water. If possible, fill the vessel with harvested rainwater from a nearby cistern or storage tank. Place a cap, lid, or small piece of tile on top of it to reduce water loss.
6. Plant seeds of garden crops with fibrous root systems—pumpkins, squashes, melons, watermelons, jack beans, horseradish, peppers, and tomatoes—roughly an inch or two away from the outside of the ollas. Immediately water and fertilize with manure tea, compost tea, or fish emulsion.
7. There is no fixed timing for refilling the olla; check the water level in the vessel frequently, and refill on demand. Similarly, replenish

An olla pottery jar, buried in compost and filled with water, is now ready to irrigate herbaceous perennial plants.

nutrients in the soil around the pot in relation to plant growth rates. Weed as needed.

8. Mark or build low barriers around the area where the pot is buried, so that humans or other animals do not trample and damage the olla.

9. Do not expect buried pitcher irrigation to be ideal for all crops or settings. Few vegetables (and virtually no grains can) accommodate the spacing it requires.

10. Seek out cheaper but more durable pots. The current cost for custom-designed ollas for buried pitcher irrigation ($16 to $30 US) makes this technique prohibitively expensive on any scale for low-income gardeners and farmers.

– Chapter Five –

Breaking the Fever

Reducing Heat Stress in Crops and Livestock

❀

Warm-Up

In the early 1980s, I had the good fortune to stumble upon what may be the hottest, driest environment in which traditional desert farmers had produced food in North America. It occurred at a place named Suvuk, on the Pinacate volcanic shield of Mexico's Gran Desierto, with at least 40 miles of lava flows, dunes, and dry *playa* lake beds between it and the nearest city. Now part of Mexico's National Biosphere Reserve for the Pinacate and surrounding Gran Desierto, it has gone as long as 36 months without any measurable precipitation; summer temperatures frequently reach 120°F (49°C). One morning in midsummer, I happened to be flying over the Pinacate just as a Mexican family began sowing corn, beans, and cucurbits in the wake of the first rains of the summer season.[1]

Curious to learn what these desert dwellers might be planting in the midst of such an extreme landscape, I drove to the edge of the lava and visited the field on foot several more times over the growing season. There, I recorded air temperatures of 114°F (46°C) in the full sun and ground temperatures of 165°F (74°C) while talking to the farmers and taking the leaf temperatures of crop plants! Rains came only two more times that year, but the silty loam in the field stored enough soil moisture from that very first rain to bring nearly all the crops—Mexican June corn, two bean species (teparies and pintos), squash, and watermelons—to their flowering stage.

As noted earlier (in chapter 2), the little desert-adapted tepary bean plants had the capacity to grow prolifically and produce beans, but nearly all the pinto beans sown the same day germinated but increasingly suffered from heat stress. The pinto plants first aborted their flowers, then their pods, and then their leaflets closed up, shriveled, and withered away. Meanwhile, the tepary bean leaflets stayed active, tracking the sun's movements most of the day. It was as if they were solar collectors programmed to expose the broadest possible surfaces of their leaves to the burning sunlight in order to capture as much energy as quickly as possible. Relying solely on three brief rains and runoff from the volcanic slopes just upstream from the field, the tepary bean plants produced a yield equivalent of 1,200 pounds per acre without any supplemental irrigation. The pintos produced less than a cup. The desert-adapted corn, squash, and watermelon varieties did nearly as well as the tepary beans.

However anecdotal such an incident may be, it reminds me of two realities that already affect our food security, and that will become even more prevalent in the future. One of these realities may bring out the pessimist in you, while the other may bring out the optimist.

The *first* reality is that many places in North America are now regularly suffering summer temperatures of greater than 100°F (38°C) whereas they rarely did so in the past. Such heat waves place unprecedented stress on crop varieties and livestock breeds that have little **thermo-tolerance** for temperatures above 95°F (35°C). When heat levels reach beyond the physiological limit of certain horticultural crops such as beans, the production of floral buds, open flowers, and pods declines, the ones that are produced abort and drop from the plant, and the number of seeds set per plant becomes drastically reduced or negligible.[2] The heat-stressed plants are also more vulnerable to leaf-eating insects and the viruses they sometimes carry. After reaching certain thresholds of heat and moisture stress—for these two factors are tightly entwined—the crop plant simply succumbs to high fever and dies.

Such plant-damaging temperatures are clearly upon us with ever-increasing frequency. In 2011, Yuma, Arizona—the closest US city to the little bean field at Suvuk, Sonora—suffered 114 days with temperatures reaching 100°F or more, and 177 days over 90°F (32°C). But such long durations of temperatures rising above 90 every day are no longer restricted to true desert regions. In 2011 and 2012, cities in the following states set new records for longest streaks of extreme temperatures (over 90°F each day) in their history: Arkansas, Georgia, Indiana, Kansas, Louisiana, New Mexico, North Carolina, Ohio, Oklahoma, and Texas. In northern states from Maine to Illinois and South Dakota, cities reached temperatures of 100°F for the first time ever, or for the first time in decades. In more southerly locales, such as those in the Hill Country near Austin, Texas, farmers had to deal for the first time ever with 90 days of triple-digit temperatures. This heat wave hit Texans just as wildfires

Date palms are used to shade house walls and reduce the heat load in Egypt's
Siwa oasis, in the heart of the Sahara Desert.

and drought wreaked havoc all around them. The following year (2012),
nearly two-thirds of the North American heartland faced comparable heat
waves and drought conditions, devastating the US corn crop and forcing
food prices to rise to unprecedented heights.

Of course, the earth's high fever has not yet broken. In a landmark
report, the Intergovernmental Panel on Climate Change (IPCC) synthesized the results of 23 climate change prediction models. Based on a set of
well-vetted assumptions about the future release of greenhouse gases,
these models predicted a globally averaged increase of between 3.2°F
(1.8°C) and 7.2°F (4.0°C) over the next century.[3] That, of course, will put
crops with low thermo-tolerance in the danger zone for heat stress
throughout most of the major food-producing regions in North America.

This brings us to the *second reality* that hit me like a revelation while I was working at the Suvuk bean field, wedged between those two lava flows in the Gran Desierto. While most crops and cropping patterns do not muster enough thermo-tolerance to help us avert a food security crisis, *some do.* It may be true that "the existing climate of much of the [Desert Southwest] is already marginal to agriculture,"[4] but innovative farmers have found ways to build "guilds" of mutually adapted crop varieties, livestock breeds, and canopy plantings, which steadily produce food even when placed in the heat of arid subtropical and tropical landscapes.

Such guilds exhibit "collective" strategies for alleviating heat stress and making the best possible use of the solar energy that cascades through their guild or micro-community. Most remarkably, farmers from all around the world have developed substantial reservoirs of knowledge about how to place these food crops in contexts where they seldom, if ever, bear the brunt of heat stress. Let's see what has worked for them that may also be adapted, refined, or used analogously in your particular foodscape.

A shade-cloth-covered Southwestern-style *ramada* allows vegetable production throughout the summer at Tucson Village Farm in Arizona.

✿

Parable

While I was spending a month in Guatemala in the early 1990s, I became aware of the fact that the concept of nurse plant guilds was well known not only in the Sonoran Desert where I was from, but in the hotter tropical reaches of Central America as well. I had spent a week trying to sleep in the sweltering heat of a multistory concrete apartment in a crowded *colonia* of a Guatemalan city before gaining access to a palm-thatched *palapa* hut in the middle of a coffee plantation. Although the two dwellings were located less than two miles apart and shared the same macroclimate, the microclimate in the coffee plantation was at least a dozen degrees cooler by noon each day. The air-conditioning there was not built into the wall of the cone-shaped hut, but came from the palm thatch of the roof, the coffee trees around the hut, and the dense tree canopy above us.

The cultivated tree that provided us such relief from the heat is known throughout Latin America by the name *madre de cacao*, mother of cocoa. It is a nitrogen-fixing legume known to scientists as *Gliricidia sepium*. It may now be the woody tropical plant most widely used across the world to provide "shade for cacao, coffee, and other shade-loving crops."[5]

Although scientists have fostered the spread of *madre de cacao* as a nurse plant for coffee to the many parts of the Americas, the Caribbean, Africa, Asia, and the Pacific Islands, its first use as a buffer against heat stress likely began in the Mayan-dominated reaches of Central America, where *madre de cacao* was first recruited to provide shade for tender young cacao plants. This must have occurred anciently, for *madre de cacao* is linked to both chocolate and arboreal monkeys in the ancient Mayan epic the Popul Vuh. As historians of the Mayan agricultural landscape have written, the cacao plant became dependent upon *madre de cacao* trees "because it requires a fine-tuned ecosystem to survive: it is wind sensitive, sun sensitive, drought sensitive, and nitrogen dependent."[6] Were it not for the tall, shade-producing, nitrogen-fixing windbreaks and nurse plants of *madre de cacao* and a close cacao relative (*balam-té*, protector tree), we might never have enjoyed the pleasure of eating chocolate or drinking hot cocoa.

In his *Travels in Central America*, first published in 1775, the Reverend Thomas Gage recalls indigenous farmers explaining to him how cacao plants needed to be protected by some kind of tree with "maternal instincts," often called step-mothers (*madrinas*) or nursemaids (*nodrizas*) in Latin American Spanish:

The [cacao] tree which doth bear this [chocolate] fruit is so deli-
cate, and the earth where it groweth is so extreme, that to keep the
tree from being consumed by the sun, they [the indigenous farm-
ers] first plant other trees which they call "mothers of the cacao"
. . . [so that] those trees already grown may shelter them, and as
mothers do, nourish, defend and shadow them from the sun.[7]

While I have been impressed by the use of nurse trees as a thermal
buffer for heat-sensitive crops like coffee and cacao in the tropics of Central America, it appears that the diversity and importance of nurse plants
is far more striking in the deserts of North and South America. In fact,
there are dozens of tree species nicknamed *nodrizas* or *madrinas* by desert
dwellers, for they are absolutely essential to the germination and survival
of a large portion of the edible flora growing in hot, dry climates.[8]

Just how much difference can a nurse plant's canopy make in protecting an understory herb or vegetable from devastating heat and damaging
solar radiation? Collaborating with my Mexican colleague and former
student Humberto Suzán, we once gathered a year and a half's worth of

Mesquite functions as a protective nurse tree for herbs at Rancho el Peñasco Eco-Lodge in Sonora, Mexico.

temperature records in the micro-environments beneath nurse trees in the Sonoran Desert. Earlier studies—undertaken well before climate change was so evident—had suggested that the dense shade of a mature nurse tree could potentially decrease the maximum soil temperature beneath it by 20°F (11°C) degrees on a summer day, and raise the minimum temperature by 5°F (3°C) on a winter day. We decided to verify those studies by measuring the soil temperature both within and beyond the understory of the desert ironwood tree, which has a dense evergreen canopy that offers continuous shade year-round.

Compared with the 115.2°F (46.2°C) temperature of desert soil fully exposed to the sun at noon, the soil temperature under the dense shade on the northern side of the ironwood canopy was only 95.8°F (35.4°C), almost 20°F degrees cooler! More remarkably, the temperature of cactus stems under the same ironwood was only 94.8°F (34.8°C), within the range at which even pinto beans could grow and flower. The microclimate under nearby mesquite trees was nearly as well buffered as those under ironwoods, with their temperatures hovering around 98.3°F (36.8°C) at noon in July.[9]

❀

Principles and Premises

One ecological principle fundamental to reducing heat stress in plants and animals is that of establishing a **boundary layer** between the sun and an organism vulnerable to excessive temperatures and damaging solar radiation. In most cases, the thermal buffer does not lie immediately on the skin of an animal or the surface of leaf, but it creates a layer of air between the "inner surface" of the organism and an "outer surface"—a leaf tree canopy, a latticework of feathers, hairs, thorns, spines, or fibers.

Think for a moment of the black robes worn by Bedouin nomads of the Sinai and Saharan deserts. Although you would at first guess that a dark robe would make a nomad in the desert hotter rather than cooler, the cloth itself is not pressed against the person's skin, but forms an air space layered between the nomad's skin and the surface of the robe, which absorbs some heat but insulates and deflects much of it away from the body. In a similar manner, black ravens, crows, vultures, and buzzards thrive in deserts, for they have a shiny latticework of feathers that reflects the sun's rays before they reach the birds' skin. Similarly, some black-skinned cattle or black-fleeced sheep create a boundary layer that keeps solar radiation from driving them toward heat stress.

Is the Desert Nurse Plant Guild Concept Really Applicable to Agriculture? You Betcha!

Just how do such ecological studies done in native desert vegetation translate to what we can achieve by incorporating nurse plants in our gardens and fields? It depends upon the crop and location, of course, but let me give you my impression from having grown greens and vegetables on the desert's edge for more than 30 years. In 2012, I grew the same variety of romaine (cos) lettuce both under the shade of desert willow (not a true willow, but *Chilopsis linearis*), and fully exposed to the sun in an open garden bed. The romaine plants grown in the open began to bolt by May 10, and their lettuce leaves dramatically diminished in size, while their bitterness grew stronger. The ones grown under a nurse plant remained harvestable until July 15, two months longer! In addition, the desert willow provided an abundance of edible flowers for our salads for well over a month. Although this can only be considered preliminary evidence, it suggests that nurse trees deserve far more attention from gardeners if we are to adapt our food production to global warming. In addition, small livestock producers would do well to keep mesquite or other legumes in the pastures, providing shade for goats, sheep, or poultry during the hottest times of the year.

A palo verde tree acting as a nurse plant for saguaro and barrel cacti in the Sonoran Desert.

A nurse tree does the same, cooling the organisms in its understory during the summer and warming them in winter by establishing its own microclimate within the boundary layer beneath the tree canopy. And yet buffering underlings from extreme temperatures is not the sole service performed by nurse plants. Since the 1930s, desert ecologists have determined that particular nurse plants like mesquite, palo verde, ironwood, hackberry, and acacia provide a wider range of benefits than just thermal buffering to plants and animals sheltered by their canopies:[10]

- Seeds are readily dispersed and accumulate beneath nurse plant canopies in soils where seedling recruitment will be favored[11] (see below).
- Canopy shade creates a boundary layer of humidity. This in turn creates a moist microclimate that buffers the underlings from death by lingering drought, catastrophic freezes, or intense heat spells.
- Organic matter under the nurse plant nurtures seedbeds with greater moisture- and nutrient-holding capacity than the surrounding desert floor, thereby fostering higher levels of germination and seedling survival.

A peach tree serves as a nurse plant, shading herbaceous perennials in its understory.

- Shallower roots of nurse plants often inoculate the seedlings of associated underlings with mycorrhizal fungi or, if both the nurse tree and the underling are legumes, with nitrogen-fixing *Rhizobia* bacteria.
- Deeper nurse tree roots often pump up or lift water, macronutrients, and trace minerals used for building the plant canopy. In addition, as they shed their leaves, branchlets, and surface roots, they deposit them in the litter below the tree, where they are composted and utilized by other plants.
- Spines, thorns, or bristles on nurse trees repel browsers and grazers that might otherwise eat or trample developing seedlings, creating prey refugia where underling plants are protected from herbivores.
- Nurse trees provide ideal nesting or roosting sites for frugivorous birds that not only carry seeds or fruits with them, but defecate them out in nitrogen-rich packages of manure.

Some nurse plants, such as honey locust and carob, function well in semi-arid temperate zones, while others provide the greatest benefits when situated in true deserts or in the arid subtropics. A few premises will help you select nurse trees for your specific locale and your own guild of food crops:

- Not all nurse plants are created equal. Some are better than others at meeting the needs of particular crops placed in their understory. Because some nurse plants are drought- or cold-deciduous, while others are evergreen, the nurse plant guilds or micro-communities beneath them are not simply an amalgam of randomly selected parts. Studies in the Mediterranean of 11 species planted under 16 different kinds of nurse trees indicate that some underlings experience different survival rates depending upon the nurse. The deciding factor for which nurse trees may be best in any case is whether the understory plant needs protection from heat, catastrophic freezes, low soil fertility, low moisture-holding capacity, or from grazing damage by herbivores.[12]
- The hotter and drier the environment, the greater the need for using older, well-established nurses with dense evergreen canopies that provide continuous shelter for underlings.
- The higher the risk of wild or domesticated herbivores browsing, trampling, or damaging the plants in the understory, the more important it is that the nurse plants selected have spines, thorns, barbs, or bristles to repel the animals.
- The poorer the soil, the more important it is that the nurse tree provides abundant leaf litter, fixes nitrogen, attracts mycorrhizae, and

pumps water to higher levels of the soil. The permaculture concept of stacking many functions into a cohesive set of plants makes abundant sense when thinking about nurse plant guilds.

Planning and Practice

Beginning around 1982, I began to imagine how we might design arid-adapted crop polycultures based on the ecological relationships found in native nurse plant guilds in the Sonoran Desert. As in other deserts,

Most tall cacti in the desert, including this saguaro, begin their lives under nurse plants such as ironwood or mesquite.

Sonoran Desert habitats feature specialized cohorts of plants that typically grow in vertical zones clustered beneath a particular shade-providing nurse tree. For instance, under a towering ironwood tree, giant saguaro cacti might grow to heights of 15 to 20 feet, with wolfberry shrubs beneath them, prickly pear cacti beneath them, and wild onions or night-blooming cereus beneath them.

After first working to understand how particular sets of Sonoran Desert species facilitated the presence of one another in these wild nurse plant guilds,[13] we turned our attention to designing agricultural systems based on certain guilds. Our most ambitious initiative was to use the nurse plant guild concept to grow perennial wild chiles and oreganos beneath the canopies of mesquites, feather trees, hackberries, and wolfberries. We helped establish one commercially producing plantation of perennial chiles under legume trees between Alamos and Navajoa in Sonora, Mexico, on the southern subtropical edge of the Sonoran Desert.[14] This field yielded an accessible harvest of fiery chiltepin peppers that campesinos mixed into their goat cheese to sell at markets as a value-added artisanal product. The feathery leaves and branches of the tepeguaje nurse trees (*Leucaena leucocephala*) were coppiced (pruned away) and used as forage for the goats that produced the cheese.

For planning and implementation purposes, the upshot of these field trials can be summarized as follows:

- Determine which combinations of understory crops and nurse trees grow best together and produce multiple benefits by evaluating natural systems nearby, or cultivated oases and other agro-forestry systems in climates comparable to your own.
- Draw a matrix of the benefits as well as the drawbacks of growing each species by itself as opposed to those of the entire guild grown together. As an experiment, grow each species by itself in a space of comparable size, and determine whether the polyculture design yields more than the cumulative harvests of the monocultures.
- Pay careful attention to establishing an appropriate height, canopy breadth, and density of the nurse plants *before* you introduce crops to the understory.
- Ensure that water is reaching the roots of both the nurse tree and its underlings.
- Remember to maintain an optimal density and size of the nurse tree canopy through periodic pruning, or else the understory crops may be completely shaded out.

- Shift your design and management, depending on whether or not small livestock like goats or free-ranging poultry like guinea fowl are present and require shade or forage. For instance, if goats are present, you can select shade-bearing plants whose foliage or seeds meet the goats' nutritional needs while being hardy enough to withstand periodic browsing or grazing.
- Measure ground temperatures within the understory and outside of the canopy on barren soil each July and December, and compare results.
- Evaluate whether or not you are gaining success in stacking functions within the same space. If you are having only limited success, redesign the spacing or the species composition of the guild.

This leguminous nurse tree provides protection from heat, freezes, drought, and wildlife damage to understory annual and perennial food plants. DRAWING BY PAUL MIROCHA.

Of course, nurse plant guilds usually take multiple years to develop, and their success in reducing heat stress is very dependent upon the landscape in which they are placed. Because of this time lag, the design practicum in this chapter will focus on the biomimicry of another kind of "boundary layer" that can be immediately accomplished around almost any kind of house or barn to produce food while reducing heat stress in your homescape. In this case, we wish to mimic the way the outer layer in a canopy of leaves helps establish a boundary layer that cools off any other object (in this case, the walls of a building) located behind that outer layer.

Up Against the Wall:
Using the Boundary Effect Principle
to Cool Your Overheated Home

At our straw bale home in Patagonia, Arizona, we have a 45-foot-long stuccoed wall for a guest room and office that is exposed to the direct sun at least six hours a day in the summer. And so, when I realized I needed a boundary layer of edible vine crops to reduce its heat load, I enlisted the help of Caleb Weaver and straw bale builder Bill Steen in designing trellises to support climbing vines. Just for fun, we decided to construct the trellises in the shapes of large (human-sized) leaves. We placed them 1 to 2 feet out from the southwest-facing walls of my home as a way to reduce summer temperatures on those stuccoed straw bale walls.

On Bill's recommendation, I purchased 8×4-foot sheets of light-gauge aluminum stock fencing that was rigid enough to serve as trellises. Caleb and I then cut them with tin snips into three shapes of leaves, and attached them by wire to the eaves of my roof. They are roughly positioned at the drip line below my gutters. We then constructed planting boxes below each of the trellises, filled each box with composted soil and mulch, and established climbing vines in the planters by seed or by transplanted cuttings. I placed them on drip lines that are minimally watered every other day, but they also receive rain off the walls and spillover from the gutters.

My first plantings were of perennial runner beans and mission grapes, but I am also experimenting with hops and passionflower vines as climbing species that may rapidly fill the latticework of the trellises with shade-producing leaves. My friend Brad Lancaster has wire trellises reaching over the south-facing windows of his home in Tucson, and uses the rapidly climbing shoots of coral vine or Mexican creeper (*Antigonon leptopus*) to provide living shade during the summer.

While people of all cultures have traditionally constructed trellises and arbors for growing shade-producing vines on south-facing walls to

Desert permaculturist Brad Lancaster uses the vines of Mexican creeper or coral vine (*Antigonon leptopus*) to shade his south-facing windows in summer, creating a cooling boundary layer.

reduce the heat loads on their homes, it is worth experimenting with shapes and spacings of trellises that more specifically use the principles of biomimicry. A good model for this modest innovation is the boundary layer effect that you can see when observing woolly, gray-green leaves of desert plants under a microscope. Leaves of many desert plants exhibit silvery pilose hairs that form a reflective boundary layer between themselves and more sun-sensitive leaf tissues. The more silvery gray the hairs appear, the more they will reflect light and re-radiate heat away from the leaves.

With a wall that is fully exposed to the summer sun, you can literally hold your hand up 2 feet away from its surface and feel the intense heat reflected off it. Once shaded by vines, however, the wall's temperature drops on the order of 10°F (6°C). Of course, these living sculptures of giant leaves are only covered with foliage in the summer and fall, but that is exactly when we need to reduce the heat loads on the walls of straw bales. In the winter, you want the wall itself to absorb as much sun as possible.

Permaculturist Greg Peterson grows edible vine crops on trellises that reduce heat loads on walls at his Urban Farm in Phoenix, Arizona.

Here is our step-by-step process for creating leaf-shaped trellises for living sculptures to develop boundary layers that reduce the temperatures of house walls or formal garden boundaries:

1. Determine the walls or orientations of your garden that gain the most heat from direct sunlight, and locate suitable sites for rectangular planters and trellises every 5 to 8 feet in front of the high-temperature surface you wish to cool.
2. Build 2-foot-wide, 4-foot-long planter boxes and place them 18 inches to 3 feet in front of the wall that is susceptible to high temperatures.

Hops form excellent vines for trellising up walls to reduce their heat loads.

We've found that 1×10 boards or scrap aluminum sheets serve well as planting box walls to enclose 8 inches or more of rich, locally generated compost. You may want to convert these planters to cold frames during cooler months, placing glass or plastic covers over them to retain heat.

3. Situate each planter box so that its "front wall" is 18 inches from the house wall, or immediately below the distance that the eaves of the house extend out past the wall. Drill an eye hook in the eaves immediately above the planter and attach a wire to it, which will later be used to stabilize the 7- to 8-foot-tall trellises that will reach up toward it.

4. Take lightweight aluminum stocking fence with 2×4-inch openings—or alternatively, build a latticework of locally grown bamboo or *carrizo* (giant cane) that is shaped into an 8-foot-high, 4- to 5-foot-wide rectangle. With bolt cutters (or pruning shears) cut the rectangle into a leaf-like image, one that is either wide and multi-lobed (like a maple leaf) or narrow and serrated (like an elm). Stand this up as trellises attached to the back wall of the planter box on the ground level, and to the wire extending down from the eye ring in the eaves on the top. The trellises should now stand 18 to 24 inches from the house wall in order to create a shaded boundary layer of sufficient buffering value in order to cool the wall itself.

5. After the last possibility of frost, sow three to five seeds of an edible vine—cucumbers, bitter melons, scarlet runner beans, pole beans, peas, passion flowers, grapes, hops—just below the trellises. When they germinate and establish seedlings, train the tendrils of the vines to climb the trellises.

Horizontal trellises for grapevines provide shade over walkways in the Gobi Desert of China.

6. As the plants mature, train and prune them to fill out the leaf-like shapes of the trellises. Occasionally measure the temperature of the house wall shaded behind the trellises and compare it with the temperature of the same wall where (and if) it is still fully exposed to the sun.

7. If you are mathematically adept, calculate the cost of building the trellises and planters versus the energy savings gained by cooling the wall instead of running a cooler or air conditioner inside the house for longer.

8. As fall progresses, remove the trellis bottom from the planter, and hinge a glazed "cold frame" cover onto the box to grow onions or greens until the following spring.

- Chapter Six -

Increasing the Moisture-Holding Capacity and Microbial Diversity of

Food-Producing Soils

Warm-Up

I had been running a "high fever," feeling sick about climate change and its devastating impact on food security, for several months before the Chefs Collaborative conference in September of 2009. But that fever finally broke when I heard Fred Kirschenmann speak to the 200-some chefs at their final luncheon at Café Brauer in Chicago's Lincoln Park Zoo. They and I trusted Fred, for he is an innovative farmer himself, having converted 3,500 acres of Kirschenmann Family Farms in South Dakota to organic grain production. Fred has also been one of the country's steadiest spokespersons for soil health and its relation to human health through his work for the Leopold Center and Stone Barns Center for Sustainable Agriculture. Whenever I had heard Fred speak in the past, I had gained fresh insights that help me with my own work at our orchard, but this time Fred opened up an entirely new range of possibilities for me. As I have tried to keep all of his words vivid in my mind, what I remember most is the matter-of-fact tone with which he spoke:

"You know, friends," Fred said candidly, "over not much longer than our own lifetimes, we've lost half of our topsoil from America's agricultural lands . . . To make up for that loss, we're using up phosphorus and potash stores for fertilizer at unprecedented rates. At the same time, we've been losing most of the soil microbial biodiversity in our food-producing ecosystems.

"But with climate change advancing, perhaps it's the careful stewardship of our soil that is the most critically important investment we can make to ensure our food security. Why? Because it's *the* foundation of everything

Without soil moisture-holding capacity in natural or artificial basins, no life can be supported.

else," he said flatly. "Both the soil fertility and water-holding capacity of our fields are largely determined by the level of organic matter in the soil, and that's what we've continued to lose at accelerating rates decade by decade ..." Then Fred came to the insight that made my jaw drop: "Look at it this way: With just 1 percent organic matter left in the topsoil of most conventional farms, a field can hold only about 33 pounds of water per cubic meter. But if you increased that organic matter to just 5 percent, the soil moisture-holding capacity goes up to 195 pounds. That's nearly a sixfold difference in a field's capacity to buffer us from the effects of a drought or a flood, and that's exactly the kind of resilience we will need to deal with climate change."[1]

Suddenly I no longer felt like I would become a victim of the drought that had hit the semi-desert grasslands around the place I was in the process of acquiring—a place in Patagonia, Arizona, that simply never greened up that season, due to the lowest level of summer rainfall recorded there in over a century. Instead, I knew, there was something I could do to build resilience, to increase the food-producing capacity of that land, however long I owned it. If the loss of soil organic matter and the carbon sequestered in it could reach 20 to 30 percent merely by taking land out of permanent perennial pasture and tilling it for annual cultivation of grain or vegetable crops,[2] couldn't the permaculture-oriented management of perennial crops on organically enriched soil rebuild that capacity? If half the water we use on our planted landscapes is lost to evaporation, runoff, or infiltration beyond the root level, aren't there ways we can immediately reduce some of those losses?[3] I tried to compose a mantra that might keep me focused on such an initiative:

> Sequester Carbon . . . Plant Perennial Fruits . . .
> Build the Soil, Harvest Water, Keep It in the Earth . . .

Adapting to climate change can clearly be done by building and sustaining the soil's organic matter levels, which can increase not only soil water content, but soil fertility as well. As R. Neil Sampson eloquently stated more than three decades ago:

> Organic matter is the single most important indicator of soil quality, and its reduction over a period of years is a sure sign that the productivity of the soil is being lost . . . The influence of organic matter in the soil is much greater than its small amounts might indicate. It serves as the "glue" that binds soil particles together into granules and crumbs that make good soil structure, keeping the soil open for the passage of air and water. Organic matter increases the amount of water a soil can hold and also increases the proportion of soil water that will be available for plant growth . . . [But] natural levels of organic matter usually drop rapidly as soon as a soil is opened up to [annual] cultivation. The reductions, on the average of about 35 percent, seem to be inevitable unless extra organic matter is consistently added to the soil.[4]

Fortunately, there is some good evidence to suggest that a diligent farmer can raise the organic matter of sandy loams from 1 to 2 percent up to 6 to 8 percent in less than a decade, and this can contribute mightily to the soil's capacity to store water, buffer itself from drought or floods, and grow nutritious food in the face of rapid climate change.[5] Starting his farm in 2003 on the edge of Metro Phoenix, my friend Ken Singh began to

A windrow of compost at Singh Farms provides nutrients and moisture to garden plantings in the urban heat island of Metro Phoenix.

mechanically rip the **caliche** layers in desert soils that had less than 0.5 percent of organic matter in their topsoil, and then added 2 to 3 feet of microbially active compost on top of them. For every 1 percent of organic matter he spread across an acre of soil, he likely added 16,500 gallons of water-holding capacity to his garden beds.[6] In less than nine years, Ken has likely achieved as much as an 8 percent increase in organic matter in his topsoil at Singh Farms, thereby increasing the moisture-holding capacity per acre of cultivated land by as much as 123,750 gallons!

Among the principles that desert permaculture activists most frequently repeat to one another is this one: "The cheapest and best place to store water is in the soil." They are not talking about high-tech means of recharging deep aquifers that have been recently depleted by unbridled groundwater pumping. They are talking about storing moisture within the soil both in and near where you are growing your food.

Soil moisture-holding capacity is a rather geeky term that simply refers to the amount of water "stored" in the soil and available for the growth of any kind of plants: annual crops, trees, perennial forages, or even weeds. When a field's soil gets drenched by rain and completely saturated by the accumulation of runoff, some of that moisture rather quickly infiltrates down below the root zone and is lost to future plant growth. The soil's moisture content at the moment when such infiltration beyond the roots ceases is known as the field capacity. Much of the remaining moisture held in the soil can potentially be used for plant growth, while other portions can evaporate or become so tightly bound to soil particles that they essentially become unavailable for the purposes of food production.

But that is when things get critical for the moisture remaining in the soil—that which is potentially available to produce food. Although a farmer wants to use some of that water for crop growth so that soil water content no longer stays at field capacity and starves the crops of oxygen, it is crucial that the soil and plant tissues do not become so depleted that the crops reach their wilting point—the moisture level at which plants can no longer extract any more water from soil. The sweet spot for plant water

The entranceway at Singh Farms is flanked by piles of compost and mounds of yam vines.

availability in between these two extremes is the soil moisture-holding capacity of arable lands suited to crop production. Within a good range of soil moisture-holding capacity figures, crop plants retain enough moisture to withstand drought, but their roots do not stand long enough in water-logged soils that they have no means to "breathe."

If soil aeration and deep infiltration are sufficient to avoid such pitfalls, farmers should aim to gain the highest possible level of moisture-holding capacity for their particular kind of soil—sandy loam, silty loam, et cetera. For a sandy loam, a farmer may want to add enough organic soil amendments to ensure that one-fifth of the soil pores—roughly 10 percent of the bulk soil volume—can still remain filled with air (not water) immediately after a heavy rain or runoff event. For heavy clay soils that are easily compacted, there is a greater risk that the soil will retain too much water after a rain, so amendments should be chosen to "lighten" or aerate the soil to the extent that is possible.

Of course, many materials can potentially be added to particular soils to increase their moisture-holding capacity without reducing aeration, but locally generated organic matter of various kinds can usually fit the bill. Prairie soils in the Midwest and Great Plains that were enriched in situ (in place) over millennia by perennial grasses and legumes once regularly harbored 4 to 6 percent organic matter in their "A" horizons, but decades of tillage, compaction, and erosion have depleted organic matter levels in some places there to only 1 to 2 percent. Today conventional farmers in the Midwestern and Plains states must spend tens of billions of dollars on soil amendments each year that are mined, manufactured, or mixed in locations hundreds or even thousands of miles distant from the fields to which they are applied.

But the dependence on external inputs for improving the soil and managing water is even more of an addiction in the arid agricultural regions of the United States. According to Ken Meter of the Crossroads Center, even some farmers in Arizona who market their produce as "locally grown" may be currently building up enormous debts and carbon footprints by purchasing their soil amendments and other inputs from distant sources. Southern Arizona farmers and ranchers spend at least $200 million each year to purchase inputs they feel they need for their food-producing operations, but gain nearly all of those purchased inputs from sources outside the region, including: fertilizers and other soil amendments; energy (mostly fossil fuel); chemicals or devices for pest and weed control; and machinery, fencing, irrigation equipment, planting stock, or feeder calves. Since they sell just $300 million of forage and food products each year, *these external inputs eat up two-thirds of the farm-gate value* of the vegetables, fruits, grains, and meat they produce. However, with $320 million of total expenditures each year to produce just $300 million of food commodities, their agricultural operations would lose $20 million a year were it not for federal subsidies and land leases.[7]

Aside from the fact that non-organic amendments seldom build moisture-holding capacity as much as they increase macronutrient levels in soils, the fossil fuel costs of petrochemical fertilizer production and transportation from extra-local sources will surely become increasingly prohibitive in the future. They may soon be priced out of reach, not only for peasant farmers in developing countries, but also for those with small- and medium-sized farms in regions of North America remote from major ports and refineries.

And yet there is also an obvious absurdity to organic farmers in the Northern Hemisphere catering to local markets while obtaining bat guano, fish emulsions, or biochar imported from South America. These amendments may indeed boost nutrient levels and increase yields temporarily, but their effects on the moisture-holding capacity, resilience, and economic viability of farming operations are underwhelming.

Fortunately, more localized sources of composts, mulches, manures, worm castings, fish emulsions, and even biochar and guano have become more readily available in many regions of North America. Today locally generated concentrations of effective microbes (predominantly bacteria) and arbuscular mycorrhizae (endotrophic fungi) cultured to stimulate root growth and moisture retention in the soil are being produced on farms or in the garden with only a modest start-up investment. Most remarkably, some sources of organic matter useful in increasing soil moisture-holding capacity can be managed to *flow on their own* into your field, orchard, or garden, and desert farmers have benefited from their root-stimulating nutrients and water-retaining qualities for centuries, if not millennia. It may be time that we replace the term *water harvesting* with *harvesting water and organic matter*, for that is exactly what indigenous and Hispanic farmers in the Sonoran Desert do to beat both heat and drought. Let's see how they have accomplished this extraordinary feat by rather ordinary means.

Parable

Through the goodwill and keen knowledge of Laura Kerman, my daughter's godmother, I first learned of the importance of collecting flood-washed detritus from the desert for renewing the fertility and moisture-holding capacity of fields and gardens. It was a practice first described among Hispanic and Native American farmers in the desert around 1894 by geographer and ethnographer W. J. McGee—a close friend of Teddy Roosevelt's—the very year that Laura was born. Laura was O'odham by birth, but she began to work with Anglo scientists, guiding them and translating their dialogues with O'odham and Hispanic farmers around Tucson by the time she was 12. She did this on behalf of her lifelong friend Dr. Robert Forbes, the first dean of the College of Agriculture at the University of Arizona. She later became a potter and clay sculptor who gathered her own clay, so that she learned much about desert soils of all kinds. But because she spent her early years working as a field guide to scientists, she remained keenly interested in all things agricultural for over nine decades. She and her brother José continued to farm a water-harvesting field into their 80s, and then "retired" to care for a sizable garden until they had nearly become centenarians.

Once, Laura asked me to take my pickup truck and gather some organic materials that José could add to their garden soil, because, as she put it, "the earth is drying up." Because I was used to gardening in desert cities, I simply went to the nearest corral, received permission to haul some horse manure away, backed up my pickup to the edge of José's and Laura's garden, and dumped out the load of manure.

José came back into the yard just as I was finishing the dumping, and became extremely animated. Because he spoke little English, he went to Laura, who was shaping some pots at the time, and asked her to translate.

"My brother says that manure is no good for this soil, it will just burn it up. I can use some of it for firing the pots, so just keep it piled up over there. But you got to go to over to the mouth of the wash where the flood-waters spread out. There, you got to get those rotting sticks and leaves and mud that flowed down from the mountains after the last rain. That's what José says will keep the soil from drying out."

I knew what materials she meant for me to gather, because I had seen huge windrows of them deposited in traditional fields in southern Arizona and all over Sonora with the summer rains, and watched as Mexican and Native American farmers plowed them under to incorporate them into their soils. At first, I had assumed it was unwanted debris—flotsam and jetsam—that simply had to be leveled and dispersed for farmers to do their work. But one of the farmers had corrected me by saying, "It is good when it comes and makes the dirt moist."

It dawned on me that what some indigenous farmers in Arizona and northern Sonora Mexico called *va:kola* or "flood-washed detritus" was

Close-up of flood-washed organic detritus to be composted for soil moisture enhancement in rain gardens.

exactly the same mix of organic materials that was deposited by the living fencerows in Sonora, where it is called *abono del río*, or "fertilizer from the river." The knowledge of how to use such nutrient-rich materials had apparently been shared among farmers (or across cultures) for centuries. I asked Laura if she had seen farmers use it much when she was still a child.

"Well, yes. When there were more [flood]waters [than there are today], it would bring the *va:kola* and white foam [phosphate-rich suds atop the water]. We could see it rolling this way [down the streambeds where the watercourses spread out just before reaching the fields]. When it gets to the [crop] plants, it covers them. Yes, my father liked it. [He'd say:] 'Don't throw it away, pile it [around the crop plants].' We would put it back in [any flood-scoured] holes as well. My father, he thought it was good for the earth in our fields."

Just what exactly makes up the flood-washed detritus? Where Laura had me collect it—at the apex of an alluvial fan on the floodplain where her people once harvested water and composted flood-washed detritus for hundreds of acres of food production—I gathered some samples in paper bags to take back to the state's agricultural college for analysis. Their most common constituent was mud-covered leaves of a nearly a dozen nitrogen-fixing legumes—mesquite, ironwood, mimosa, three kinds of palo verde,

Building Soil Means Building Moisture: Trapping Silt and Organic Matter Along Dry Washes in Sudan

I would be remiss if I left you with the impression that only indigenous farmers in the Sonoran Desert have found effective means to harvest flood-washed soil and composted organic matter along with rainwater. In the Red Sea Hills of northern Sudan, farmers of the Beja culture have accomplished virtually the same process of soil renewal and water retention. They build silt traps in and near the millet and sorghum fields placed along the usually dry *wadi* or *khor* streambeds in narrow valleys that are naturally vegetated by two acacias known as samer and sayal. These two nitrogen-fixing legumes not only protect the streambanks from erosion, but generate the organic matter that gently floods into Beja fields after being slowed by the careful placement of low-lying crescent-shaped embankments immediately above the grain fields. Not only is water delivered to the crops, but silt and organic matter are deposited in a manner that renews and improves soil texture and moisture-holding capacity.[8]

Of course, the Beja culture is not the only community in the Old World to harvest organic matter along with the stormwaters that occasional course down ephemeral waterways. It is presumed that the ancient Nabatean culture of the Negev Desert in present-day Jordan, Palestine, and Israel had similar means of capturing and utilizing flood-washed detritus in their fields and orchards. The acacias and other shrubs that sparsely cover the upland soils of the Negev were undoubtedly allies to the Nabateans in successfully building soil moisture-holding capacity in their arid region between the Dead Sea and the Red Sea.

and several acacias. Next, there were branches and twigs and flower spikes from these same legume trees, including charred wood with many of the characteristics of biochar. (In chapter 8, I will describe how the prehistoric Hohokam culture of the Sonoran Desert cultivated thousands of acres of agaves or century plants for food and beverage using materials quite similar to biochar. They incorporated into their desert soils tons of charcoal made from woody legumes from their roasting pits.[9])

The third component of the detritus surprised me, for it is widely dispersed on the desert floor, but seldom seen in any abundance unless the floodwaters collect it into concentrated masses: cottontail droppings, jackrabbit droppings, rodent droppings, and peccary (javelina or "wild pig") poop. We could also identify seeds and fruits from at least seven desert trees and shrubs, but most had a fine sheen of mud covering them.

Of course, buried in wet mud on a 100°F day, most of these organic items did not keep their shape or consistency for long. They were rapidly composted, or, as the desert farmers describe it in their vernacular English, they "rotted" or fermented. Another farmer and friend of Laura's explained flood wash to me in this way: "It is all the different kinds of things [in the desert] that get rotten and break down whenever the rain comes and washes them away [into our fields]."

Listening to these farmers during many conversations at my home and theirs, it seemed that they viewed this act of fermentation or nutrient renewal triggered by the summer rains to be key to desert renewal. Let's explore what lies beneath this seemingly simple process practiced by many desert farming cultures around the world.

❀

Principles and Premises

Although not everyone has access to local organic matter as useful as "pre-composted" flood-washed detritus for increasing soil moisture-holding capacity, there are other locally available organic materials that may be nearly as valuable. In many dry regions around the world, enormous quantities of organic materials are available, but remain underutilized. Where I live near the bordertown of Nogales, Arizona, an estimated nine million pounds of compostable produce waste is deposited in landfills each year, although efforts are now underway to rescue much of it for use in making soil amendments. School cafeteria wastes, tree prunings, lawn clippings, livestock manures, cottonseed meal, feathers from processed poultry, bloodmeal, bone meal, ground-up hooves and horns, storm sewage sludge, algae from stock ponds, organics from sewage treatment facilities, apple pomace, and mold-spoiled hay may be available in considerable quantities in your area. To the extent possible, avoid organic solids from sewage that are likely to contain pesticides, herbicides, and antibiotics, for as Ken Singh reminds us, your compost will ultimately only be as good (or as rich in microbes) as the feedstock with which you nourish it. The nutrient composition and toxic load of most of these materials are well known,[10] and materials deficient in one kind of nutrient can be blended and mixed with other materials rich in that same nutrient.

However, finding comparisons of the relative moisture-holding capacity of different soil amendments is more difficult, so it is best to directly monitor your particular blend of amendments when fully mixed into the parent soil materials on your site. In arid and semi-arid regions, be particularly careful not to use too much animal manures having high salt content.

The principles for increasing soil moisture-holding capacity are relatively straightforward:

A compost heap is placed in paneled boxes on the edge of terraces where both annuals and perennials grow.

- The closer the source of compostable organic matter to your site of food production, the more cost-effective it is to frequently gather it and massively process it.
- The greater the organic matter content in the root zones of your food crops, the more rapid the growth of those root systems and the lesser the risk that they will not retain enough moisture to avert drought.
- The drier the climate where you live, the more frequently the organic materials need to be watered, layered, and turned in order to make compost into effective, stable humus over a shorter period of time.
- Because organic compost comprises more than just macronutrients such as nitrogen or potassium, the broader the range of materials mixed into the compost heap, the more varied and balanced its nutrient composition is likely to be.

Because of His Deeply Composted Soils, Ken's Vegetables Really Singh!

Singh Farms produces both compost and vegetables on the Salt River Indian Reservation in Metro Phoenix.

My friend Ken Singh is Arizona's acknowledged master at microbially transforming high-quality local "feedstock" materials into highly fertile moisture-retaining soils in the Sonoran Desert. Ken weekly composts tons of desert tree prunings, herbicide-free clippings, and other truly organic materials that come from within 20 miles of his 20-acre site on the Salt River Indian Reservation near Scottsdale, Arizona. Inoculating the newly arrived materials with the effective microbes in hydrolized fish emulsion—and benefiting from the high temperatures of central Arizona for "cooking" the compost—Singh completely transforms 10- to 12-foot windrows of organic matter in a matter of five to six months. The final material is so rich with mycorrhizal fungi that you can see their white hyphae in every handful you pick up out of Ken's matured compost piles. Ken then adds 2- to 3-foot-deep layers of it to the dozen or so acres of tree crops and vegetables he grows in the hottest reaches of the Sonoran Desert. Some of the fruit trees he planted as foot-high saplings

in 2005 were already 20 to 25 feet tall by the beginning of 2013, and they now perennially bear abundant fruit crops. In the shade of those trees, the salad greens and vegetables he grows in the mineral-rich compost have such high Brix levels, or soluble sugar contents, that their flavor and texture are incomparable. Once Ken offered me a large dark red leaf of Swiss chard with thick crimson veins running through it, and laughingly warned me, "Be careful when you bite into it! You may think that you are eating some grass-fed meat!"

Singh essentially gains the same quality of rich organic matter that Native and Hispanic Americans once achieved from having it naturally funnel down the intermittent desert streams above their fields. Whether you gather organic matter and nutrients passively, as happened historically in the Sonoran Desert, or actively, as Ken Singh does with a dump truck today, gather it in abundance and compost it with both the effective microbes and fervor that are manifested on Singh Farms!

A cobble-edged micro-catchment harvesting rainwater at Trincheras, Sonora, Mexico.

- The hotter the compost heap, the faster it will cook, but the greater the volatization and de-nitrification rates of nitrogenous materials are likely to be, unless microbes fix them into organic forms.
- The sandier the soil, the greater its capacity to absorb additional organic matter.
- The more clay in the soil, the greater the need there will be for sulfur or gypsum (calcium sulfate) amendments to improve the nutrient- and moisture-holding capacity of the soil.

At the University of Arizona's Compost Cats Project, composting to improve the moisture-holding capacity of soils is being done on as massive a scale as possible. PHOTO BY CHET PHILLIPS.

- The longer the frost-free growing season is for food crops, the greater the need to periodically replenish the organic matter in the soil with substantial dressings.
- The higher the salinity or alkalinity of the material (such as horse manure), the more important it is to mix it with other materials and keep it longer and wetter so that the salts may be leached away.

❁

Planning and Practice:
Capturing Compostable Flood-Washed Detritus

Of course, the first step in determining where you should build greater soil moisture-holding capacity is checking the current status of moisture in your garden or field, not just at one spot, but at many. There is a simple, hands-on way to judge the current level of moisture in your foodscape, just as soil scientists in the Natural Resource Conservation Service have done for years (table 6-1).

Once you've determined the soil texture and current moisture level at different spots in your field, orchard, or garden, you can better determine the amount and kinds of amendments that best fit their needs, and monitor soil moisture changes to avoid plant stress. If soil moisture levels in the root zone drop below 15 percent in clayey soils or 5 percent in sandy soils, most annual crops will be likely to wilt.[11] If soil moisture levels do not rise above 50 percent in the root zone of trees and other woody perennials, they, too, will suffer drought stress. Of course, any soil that stays waterlogged at or near field capacity for several days after a downpour or flood is likely to starve the roots of perennials of the oxygen they require to stay alive.

Of the many amendments we can add to mitigate floods or drought stress in the soils of our gardens, fields, and orchards, the most neglected are the ones found closest to home. This includes the extraordinary proportion (14 to 33 percent, by some measures) of all the foodstuffs and their trimmings that reach our kitchens before they spoil or are abandoned for use and thrown away.[12] However, when we look at the wastage of all organic biomass from the time a crop is harvested in our gardens or fields to the time a portion of its volume lands in our mouths, closer to 50 percent of the edible yields we grow are wasted.[13] It takes

TABLE 6-1 Guide for Judging Soil Moisture

ROUGH % OF RESIDUAL SOIL WATER	FEEL IN LIGHT-TEXTURED SOIL	FEEL IN MEDIUM-TEXTURED SOIL	FEEL IN HEAVY-TEXTURED SOIL
0–25	Dry, loose, falls between fingers	Powdery dry. Often crusted but easily crumbled into powder	Hard-backed and cracked, with few loose crumbs on the surface
25–50	Appears so dry that it will not form a ball	Rather crumbly but will hold together as a ball under pressure	Somewhat pliably shaped into a ball under light pressure
50–75	Tends to initially ball under pressure, but doesn't stay held together	Plastically forms a ball that may form a slickly sealed surface under pressure	Quickly forms a ball that can extrude into a ribbon between thumb and finger
75–100 (field capacity)	Forms a weak ball that breaks down and will not slick	Forms a very pliable ball, one that slicks readily	Quickly forms a slick ball that then squirts out between fingers
At field capacity	Upon squeezing, no water extrudes, but ball leaves wet outline of hand	Upon squeezing, no water extrudes, but ball leaves wet outline of hand	Upon squeezing, no water extrudes, but ball leaves wet outline of hand
Oversaturated	Water droplets form on ball and hand	Water droplets form on ball and hand	Water droplets form on ball and hand

Adapted from SCS Bulletin #199

both planning and persistence, but much of this produce can either be rescued for direct consumption by humans, their poultry and livestock, or used as a matrix for generating compost, worm castings, mulch, and other soil amendments.

Even if we divert more of the food flowing through our land to our kitchens into soil amendments, there remain other sources of organic matter that may be deposited on our lands or in our waters but that are seldom captured and transformed for enhancement of our soil's moisture-holding capacity (table 6-2).

These materials may vary widely in quality and quantity of nutrients available from place to place, but they deserve far more consideration and utilization than they have received in the recent past. Particularly where these organic materials accumulate on your own land or in public spaces nearby, you can rescue them from being dismissed as a nuisance or disposed of as trash, and provide nutrients as well as enhance the moisture-holding capacity of your soils. You can now use the Compost Calculator app for Androids to select from 25 different raw materials for composting, and determine how much of each you need to achieve optimum carbon-nitrogen ratios.

Make Bokashi, Not War! Developing Compost Trenches for Orchard Crops

Because I have *terra rosa* clay soils across most of my orchard, tree waterlogging and death by oxygen starvation were among the first problems I faced when I began to plant fruit and nut trees. I've had to resort to renting a jackhammer with a shovel attachment to excavate trenches in the hard-packed clay on the ridge above our home. But once I have opened up a yard-deep, yard-wide trench upslope of the tree crowns of my heirloom apples, olives, pawpaws, and persimmons, I can use the **bokashi** method of rapid and intensive fermentation of organic matter with effective microbes to enrich the soils with a loose, almost porous material that reduces the likelihood of root waterlogging.

The bokashi fermentation method originated in Japan, but it has already spread into many parts of Asia, Europe, and North America. It has been incorporated into a traditional deep-ditch manuring process in arid regions of China to increase soil nutrients and stabilize moisture levels in tree root zones, thereby increasing the length and density of small-diameter functional roots on apple trees.[14] But whether you do bokashi composting in pre-manufactured bins or in deep trenches, the key is to inoculate and activate the compost with effective microorganisms such as lactic acid and phototropic bacteria, fungi, and yeasts. Anything but cardboard and bones can be heaped into the trench, including cooked meats and curdled dairy products, as well as other kitchen wastes. Once inoculated with a liquefied EM (effective microorganism) culture, the organic materials tossed in the trench can be ready in as little as three weeks as long as they are kept moist.[15]

I dug my yard-deep trench and began to inoculate and water kitchen scraps, yard prunings, and horse manure about two weeks prior to the arrival of saplings of six varieties of American persimmons and pawpaws. I planted the trees immediately adjacent to and downslope from the trench, so that most of the roots would be encouraged to grow "upslope" rather than toward the clay-dominated reaches of the row where most of the runoff and rain will accumulate after a thunderstorm. A yard-wide catchment basin was shaped around each fruit tree and "sand-bagged" on its downslope side as a provisional terracing method while the trees were getting established. Later, when the trees are well rooted, I will remove the sand-filled burlap bags and dig a second trench on the downslope side of the tree crowns.

The flood-washed detritus or *abono del río* referred to earlier in this chapter is an example of the valuable materials that most of us neglect to rescue and restore for use in our food system. What we need to initiate, then, is a more comprehensive audit of the kinds of sponge-like organic matter that enter our food system that we can potentially divert to build our soils' moisture-holding capacity:

1. Walk across your food-producing landscape and the watershed up-stream from it, and flag with tape all the places (stream meanders, tree canopies, fences) where flood-washed detritus naturally accumulates after storms.

2. Number each deposition site, then estimate the depth and diameter of accumulated organic matter, and its distance from your field or garden.

3. Rank each site on the relative ease of access (by wheelbarrow, wagon, or pickup truck) to its organic deposits.

4. Gather a wheelbarrow load of material from each kind of deposit, and make an open, "hot" compost heap with it. Water it once, add effective microbes if available, and then evaluate over the next month its temperature, texture, and degree of decomposition. Select the sources with the greatest volume and speed of composting.

5. Determine which of your cultivated soils require the greatest volume of moisture-holding soil amendments, and match the kinds of organic materials you have available to each soil type.

6. Amend each soil with a 1:1 ratio of compost in the top 6 inches.

As suggested above, once you have inventoried these flows of organic matter, you may wish to determine whether it is more cost-effective to blend and process them, or to use each of them separately for a different purpose. The time, energy, and space required to process each kind of material may vary greatly, but the use of effective microbes, worms, and other invertebrates to speed up the processing time also deserves consideration.

The construction of brush weirs or fredges are means that desert farmers worldwide have used to accumulate flood-washed detritus as a compostable amendment. Although brush weirs only act as effective silt traps (for collecting soil, leaf litter, twigs, stems, and animal manure) in certain geomorphic settings along watercourses, it is nevertheless possible to employ this passive means of accumulating compost in many more contexts than where it is currently done. Here are the steps for doing so.

1. Locate the place along a watercourse or hillside drainage above (upstream from) a field where tributaries are no longer coalescing but have reached a transition in gradient where they are beginning to braid apart to form a flatter alluvial fan. This is what is traditionally called an alluvial delta apex or "the mouth of a wash."

2. Set a spirit level, water balance, or board across the delta so that it runs perpendicular to the prevailing flow of water; establish a line transecting the delta; and dig 4-inch postholes every foot along this line, with each hole 8 to 12 inches deep.

3. Anchor 3-foot-tall stakes of willow, cottonwood, alder, mesquite, or acacia in these holes, so that the stakes stick up at least 2 feet above ground level. Weave bamboo stalks, "carrizo" cane reed, willow

TABLE 6-2 Organic Materials Available
on Farms and Ranches for Composting in Most Hot, Dry Climates

MATERIAL	VALUE	RELATIVE SCARCITY	PROCESSING COST
Table scraps and trimmings from vegetable harvesting and processing	Variable in nutrients, higher water content, excellent in hot compost	Available in abundance from cafeterias and cafés, but volume per day is lower from homes	Can be hot-composted at low cost, but requires periodic watering
Flood-washed detritus in streambeds and floodplains	Variable, but typically rich in nutrients, excellent for compost	Relatively spotty in distribution; sometimes hard to access from muddy streambeds	Because it is pre-composted by stream or sheet flows, virtually no cost other than gathering
Native tree bark and branches run through chipper-shredder	Low in nutrients, high in carbon, good as mulch or for loosening clay soils	Typically more available than locally utilized	Can be composted at a low cost, but quicker turn-around time if watered
Native tree leaf and twig litter with small mammal droppings	Relatively high in nutrients and weed seed, partially pre-composted; used for amending sand and clay	Locally abundant, but widely dispersed in small volumes	Gathering can be time-intensive, but hot-composting to kill weed seeds is relatively cheap
Manure, blood, and bone-meal from large livestock (especially if animals were not given antibiotics)	High in moisture content (66–86%) and potash (17–34%) but low in nitrogen (0.7–0.5%) and phosphate (0.02–0.3%)	Locally abundant in rural areas; easy to obtain, but can be laced with salts or agri-chemicals	Low processing costs (though 3 years needed to leach out antibiotics)
Manure and feathers from poultry	Lower in moisture content (55–75%) but high in organics (25–45%), nitrogen (0.6–1.1%), phosphate (0.8–1.4%), and potash (5%)	Increasingly accessible in urban as well as rural settings	Minimum processing costs or time delays

University of Arizona students known as "Compost Cats" collect organic materials from campus and nearby businesses to increase compost availability for school gardens in Tucson, Arizona. PHOTO BY CHET PHILLIPS.

osiers, Osage orange branches, or elderberry shoots horizontally between these stakes, just as in the construction of a living fencerow or fredge (see chapter 2).

4. Clear away obstructions in the watercourse just upstream (1 to 30 yards) from the weir.

5. After each rain that generates runoff, take a wheelbarrow and shovel to the weir and haul away the flood-washed organic detritus that was deposited there. If not fully composted, add to your compost pile; otherwise, apply 6 inches deep across the portions of your cropland with the lowest moisture-holding capacity.

While larger, legume-dominated watersheds will likely produce the most nutrient-rich compostable detritus, this technique should be integrated into water-harvesting designs wherever possible. Remember: When rainwater and nutrients come together in the same seasonal pulse, the two most critically limiting resources for plant growth are made available at the same time, and a substantial harvest of nutritious food is the likely reward.

These tree trunks block and accumulate compostable food wash.

Chapter Seven

Forming a Fruit and Nut Guild

That Can Take the Heat

Warm-Up

I did not begin to grow an orchard of 60 heirloom fruit and nut tree varieties in order to resist heat or sequester carbon. Instead, I sought to emulate the multiple strata of woody perennials I had seen planted in desert oases in Egypt, Morocco, Tajikistan, Sonora, and Baja California for hedonistic reasons: to reap their delicious fruits and nuts for eating fresh, drying, and canning. But when I learned how industrialized agriculture was using more calories of fossil fuels in growing annual monocultures than the caloric yield of nutritious foods being harvested from those croplands, I realized that this earth needed as much perennial cover as our efforts in planting trees could muster. Because the Intergovernmental Panel on Climate Change has estimated that annual carbon emissions are likely to triple worldwide by 2100, approaching 20 billion metric tons per year, scientists have even proposed the construction of "synthetic trees" to pluck carbon out of the air in order to slow global warming.[1]

And yet, as I remember nearly every time I go out at dawn to care for my orchard, there is a certain magic associated with planting, tending, and harvesting fruits from real trees that encourages us to engage with them as often as possible. In 1953, when the great French novelist Jean Giono wrote his classic about the tree-planting nomad Elzeard Bouffier, *The Man Who Planted Trees*, it was rejected by *The Reader's Digest* as a preposterous fable about an imaginary character. But as soon as it was published in *Vogue* in 1954, it created a firestorm of enthusiasm for planting woody perennials. Since that time, it has been translated into dozens of languages, inspiring tens of millions of people to participate in tree plantings in both urban and rural settings. In the words of my own writing teacher, poet William Stafford, it became "a story that could be true." As literary historian Tamsin Kerr has playfully suggested,

> Elzeard Bouffier has had a posthumous existence as remarkable as his imaginary real life. He has gone and planted his acorns and grown his forests all over the world, from New Zealand to Kenya, from Finland to the United States. Whenever his story has been published [in another language], people have believed in it.[2]

It may seem like a rather small, humble act to plant a single tree, but even one dwarf or semi-dwarf fruit tree sequesters an enormous amount of carbon over its life span. A dwarf fruit tree may sequester as much as 28 pounds of CO_2 a year, while a larger semi-dwarf or full-sized tree will sequester between 220 and 260 pounds annually. Their **carbon sequestration** into their root mass and surrounding soil may peak in about 20 years, but fix in place as much as 5,000 pounds of carbon over its lifespan. That is no small feat.[3]

Plant an acre of apples or other longer-lived trees in an orchard, and you will have done a great service not only to humankind, but to all life on earth. As Cornell University researcher Alan Lakso has calculated for an orchard in Geneva, New York, several positive effects on our climate accrue from each such planting. An acre-sized apple orchard can fix as much as 20 tons of CO_2 from the air each season, while it releases 15 tons of oxygen, and provides over 5 billion BTUs of cooling power.[4]

That is not to say that annual crops have no value, but vegetables tend to fix only 9 or 10 grams of carbon per square meter each year, while flood-irrigated rice sequesters five to six times that. But when the use of arable land in California has switched from annual crops to perennial vineyards, it sequesters as much as 68 grams per square meter each year. When annual cropland is converted to orchards, it sequesters up to 85 grams per square meter per year.[5] Roughly speaking, vineyards sequester 7.5 times more carbon than annual crops, and orchards sequester 9.4 times more than annual crops. If a second strata of mixed perennials and annuals were planted in the understory of taller fruit and nut trees in an orchard, that land could easily sequester 10 to 15 times more carbon than most annual monocultures. Overall, altering the mix of food varieties and species to perennial ones may be one of the most critically important shifts we can make if we are to adapt to *severe* climate change, since most other

Diagonally espaliered pears work well in temperate zones, but may or may not receive their required winter chill hours in arid, semi-arid, or subtropical climates.

arable land management options will only work if the degree of climate change is low or moderate.[6]

For all of you wanting to mitigate climate change through tree plantings, the carbon sequestration value of fruit and nut trees is as important as their value in cooling nearby buildings, in holding soil in place, and in creating favorable micro-environments for understory crops. The bad news is that we may not be able to grow the same varieties of fruits and nuts in our vineyards and orchards as we have in the past, as a result of the dramatic reduction in chill hours and dramatic increases in maximum summer temperatures that have become commonplace with global warming. Let's learn more about the challenges of producing fruit and nut trees under rapid climate change from farmers who have already been facing these challenges.

Milder Winters, Earlier Springs, Hotter Summers, Fruitless Falls?

Many of the temperate deciduous tree species that produce fruits or nuts require a chilling period in the winter and no extremely hot temperatures during flowering in order to produce. Different species and even varieties of trees require a minimum number of chill hours to be fruitful, but the physiological mechanisms that set this so-called requirement vary from species to species. As a rule of thumb for selecting varieties to put in at a site, arboriculturists count the number of hours in the dormant season (from November to February in the Northern Hemisphere) that a site experiences temperatures in the range of 32 to 45°F (0–7°C), and matches up the site's accumulated chill hours in recent years with nurserymen's rough estimates of the number of chill hours "required" by certain fruiting varieties. If the tree fails to receive the "required" number of chill hours over any cool season, the following problems may arise: irregular breakage of dormancy and spotty bud formation; tardy leaf development; delayed or sporadic blooming; asynchrony of blooming with pollinator availability; reduced fruit set; and compromised fruit quality.

Although particular grape varieties may have a chill requirement ranging anywhere from 100 to 600 hours, the quality and quantity of the fruit produced are more dramatically (and adversely) affected by high summer temperatures. If nighttime temperatures pass a certain threshold—one that may be different for each variety—the fruit matures too quickly and develops too much sugar, but the flavor components develop too slowly. The possible exception to this is in red and purple grapes, which accumulate flavor-rich phenolic compounds during summer heat and drought stress. To a large degree, wine prices are negatively correlated with high summer temperatures. The hotter the summer, the more syrupy dessert wines and brandies arrive into the marketplace.

Nuts such as pecans are also highly sensitive to hot summer temperatures and high humidity. Their nuts may literally attempt to germinate in the shell on a hot, humid September day, leaving little to harvest or to sell.

Parable

I had not exactly fathomed how much traditional orchard-keepers were already being impacted by climate change until I met two fruit growers in the Pamirs of Tajikistan, one of the driest highland regions of Central Asia. Wedged in between the Tien Shan range to the east and the Himalayas to the south, the agricultural valleys of the Pamirs receive only 10.5 inches of rainfall in an average year. And so, over the centuries, indigenous farmers in several steep valleys there have mastered the construction of canals reaching up above 16,000 feet in elevation to where water from the summer melting of glaciers can be transported miles downstream to irrigate crops where the land flattens out into narrow floodplains. I had the opportunity to join a field trip into the Khuf Valley of the Pamirs, where Rushani-speaking farmers were not only competent managers of high-mountain irrigation systems, but were experts at growing grain, fruits, and nuts there as well.

As we started our 10,000-foot climb from the roaring River Panj at a 6,000-foot elevation, our Russian-built van lunged and jerked in low gear up the steep rocky road nestled in a nearly barren canyon. But 3 miles up the road toward the village of Khuf, the van overheated, and six of us got out to climb the zigzag of a dozen switchbacks until the vehicle and its Tajik driver could catch up. Fortunately, this unanticipated meander on foot allowed us to spot several acres of greenery ahead of us, where an orchard was situated in a spot surrounded in every direction by large boulders and loose scree slopes.

When we arrived at the orchard's entrance, we were greeted by two elderly Rushani farmers who offered us some of the fresh fruit they had just harvested, along with some tea they had brewed moments before. The older of the two men then explained to us why and how they had come to plant this and a number of other orchards at different elevations, as a Tajik botanist, Ogonazar Aknazarov, translated their story for us.[7]

As lifelong residents of the Khuf Valley, these men had become concerned several years ago that the sites where their ancestors had planted traditional varieties of fruit and nut trees were no longer producing. They were not sure whether the sites were depleted of their nutrients and water-holding capacity, whether soil diseases and insects had accumulated to intolerable levels, or whether the old varieties no longer functioned well physiologically under present conditions. And so they decided to plant a mix of old as well newly introduced varieties at a number of "virgin sites" scattered along an elevation gradient like so many beads on a necklace. Just five or six years earlier, the elderly man had first begun his experiment, planting over 100 trees representing more than several dozen varieties at the very orchard within which we sat. He planted others like it at 500-foot elevation intervals both above and below his ancestral village of Khuf, to see if some would produce more nuts or fruits than they did in the old orchards immediately surrounding the village.

Because we knew that climate change had triggered accelerated melting of the glaciers at the top of the valley, we asked the elder whether the entire valley was getting warmer. His answer puzzled me at first, until I better understood the airflows between the glaciers above us and broader River Panj floodplain below us.

"Well, it depends where you are [on the gradient]," the elderly orchard-keeper explained to Oganazar, who then translated. "At the very

These walnuts in the Khuf Valley of Tajikistan can no longer produce in areas where they have been grown for centuries, because of reduced winter chill hours.

bottom of the valley, where our stream flows into the Panj, it is far colder in the summer than it was when we were children. We used go sleep outside in the summer there. But now it is too chilly at night. Some crops mature more slowly there than they used to . . . Others, like mulberry, have become harder to grow."

He then went on to suggest that the reverse was true at higher elevations. The frost-free growing season had become longer, with winter and spring warmer than they had been at any point they could remember over their lifetimes. The current summer temperatures were not unusually high compared with other parts of Central Asia, but the melting of the glaciers and the frequency of mudslides below them were unprecedented. Farmers had begun opening up higher pastures for their Karakul sheep, and

136

planted wheat, barley, potatoes, and corn hundreds of feet higher than they had done historically when they had been more limited by severe winters. Meanwhile, the older trees of the fruit and nut varieties planted by their fathers and grandfathers were no longer leafing out, or flowering and fruiting with much vigor. Their yields had trickled down to a minimum, and the quality of the few mulberry fruits or walnuts produced was compromised.

As I listened to the translations of these farmers' words, I realized that their trees were suffering from a reduction in chill hours, which is an indication of their need for cool winter temperatures between 32 and 45°F (0–7°C) that trigger budding, flowering, and fruiting in the warm season. Without such a trigger, some species of temperate trees do not develop enough fruit to make them worth harvesting. Most temperate fruit tree varieties require 750 to 1,750 hours of cool (32–45°F) temperatures during their dormant season to produce a crop the next year. When climatologists David Baldocchi and Simon Wong attempted to predict how fruit production in the year 2100 will be affected by the climate changes occurring since 1950, they forecasted an average loss of 500 chill hours, roughly half of what most temperate fruit species require! They found that from 1950 to the present, various fruit-growing localities have lost from 31 to 161 chill hours—a reduction of roughly 5 to 26 hours per decade.[8]

A more recent analysis used several different emissions scenarios and global warming curves, but still concluded that the "safety zone" sufficient for adequate winter chill required by many fruit tree species or varieties will decrease 50 to 75 percent by the mid-21st century, and 90 to 100 percent by the late century. While many of the fruit tree varieties grown in Central California will not be beyond their safety zone for winter chill until 2025, it appears that the time-tried varieties of the Khuf Valley in Tajikistan may already be beyond their threshold.[9]

While the Rushani-speaking farmers certainly did not know the chill hour science behind their observations, they clearly understood that they could either shift their sites for fruit tree production along an elevational/temperature gradient, or they could shift varieties upward or downward along the same gradient. Rather brilliantly, it appears that a few Rushani-speaking orchard-keepers were attempting both strategies on a modest "wait-watch-and-see" scale. They were eager to share their results with their neighbors, who could then use the climate change adaptation strategy called **assisted migration** to move their varieties down the valley toward the River Pani, or upstream, toward the glaciers. If they continue to do so with success, they will likely be implementing one of the most effective uses of agricultural biodiversity in adapting to climate change.[10]

In Granada, Spain, small orchard gardens act as private agricultural experiment stations called *almunias*, where fruit crop varieties such as those of pomegranates are evaluated.

Principles and Premises

In principle, we have much to gain from increasing the portion of agricultural lands planted to perennial crops, as long as we select species and varieties that will have their optimum temperature needs still being met 15 to 30 years after planting. Here are the ways in which planting perennial vines, shrubs, succulent rosettes, and trees to produce edible fruits, nuts, pods, or tubers can also help us adapt to climate change:

- Their roots, trunks, branches, and foliage can sequester more carbon per year than most annual crops do, and their sequestration rates per year tend to increase over time, peaking for many trees in the third decade after planting.
- The taller, bushier (more densely leafed) trees, shrubs, and palms may also reduce heat loads on understory plants or on the walls of nearby structures, thereby allowing the carbon emissions from coolers and air conditioners to be reduced.
- In addition to creating their own microclimates, trees and palms also provide trellising and support structures for vines and other plants with scandent (sprawling) habits.

- Many (but by no means *all*) perennials can tolerate prolonged drought, floods, and wind to a greater degree than temperate and tropical annuals can.

With regard to the last point, British horticultural designer Alice Bowe has articulated an interesting strategy for selecting perennial plants to weather climatic uncertainty. She suggests that we search for perennials that are **bimodal** in their capacity to withstand both drought and floods:

> The changeable weather we have been enduring has highlighted the need for a new approach to our planting choices. In previous years we would label gardens [or specific species] as "reliably moist" or "subject to drought" with some ease and then plant them

A leaf-lined runoff drainage way irrigates a fruit tree during summer rains in Terrenate, Sonora, Mexico.

TABLE 7-1 How to Identify Plants Tolerant of Four Climatic Stresses

CLIMATIC STRESS	DROUGHT/ HEAT	WIND/ STORMS	FLOOD/ WATERLOGGING	FIRE/ BURNING
Foliage/branch structure	High surface-to-volume ratio, with succulence in some	Flexible, wispy, quick to shed damaged outer branches	Multi-stemmed from the base (in case some die)	Vertical, with basal offshoots rapidly rising after a fire
Bark/skin	Thick, glossy, gummy	Pliant (elastic or resistant to breakage)	Slick, smooth	Thick, fire-tolerant
Leaf color	Gray, silver, or reflective	—	—	Oily green or gray
Leaf surface texture	Hairy or woolly, creating a cooling boundary layer	Thick, waxy	Quick drying	Laden with volatile oils
Leaf shape/size	Small or dissected	Needle-like	More often linear than widely lobed	
Trunk/stem	Swollen to hold moisture inside	Low center of gravity, stout	Multi-stemmed, capable of resprouting	Multi-stemmed, capable of resprouting
Root shape	Either deep or shallow, with many lateral or one swollen unit(s)		Lateral; tolerant of being buried or exposed, some with breathing tubes	Rhizomatous; capable of laterally resprouting

Orchards are surrounded by shelterbelts and hedges in rain-shadow habitats in Afghanistan, buffering fruit and nut trees from winds and floods.

accordingly, but things are becoming less and less predictable . . . For the unpredictable weather challenges ahead, we will need to develop a sophisticated arsenal of bimodal plants—that is, plants that can cope with more than one type of weather extreme . . .[11]

Although the species considered tolerant of drought, wind, flood, or fire stresses may vary wildly from place to place, there are observable cues in the architecture, coloring, and adornment of plants that might guide us in discerning where their tolerances may lie (see table 7-1). In this case, we might term Alice Bowe's concept **quadri-modal**.

In addition to seeking out fruit and nut trees tolerant of these four stresses, you will also want to select for planting those that require far fewer (100 to 200 fewer) chill hours than what your foodscape is currently experiencing. A provisional list of low-chill fruit and nut varieties (excluding citrus and other tropical species) is provided in table 7-2.

TABLE 7-2 Low-Chill Fruit and Nut Varieties Suited to Hotter, Drier Climates
with Plant Hardiness Zone suitability and self-fruitful (sf) capacities noted

COMMON NAME/VARIETY	SCIENTIFIC NAME	CHILL HOUR RANGE	OTHER ADAPTATIONS
Almonds	*Prunus dulcis*		
All-in-One		300–500	Tolerant of Zones 7–9, sf
Garden Price Dwarf		250	Hardy in Zone 9, sf
Ne Plus Ultra		250	Tolerant of Zones 7–9, sf
Nonpareil		400	Tolerant of Zones 7–9, sf
Price		400	Tolerant of Zones 7–9, sf
Apples	*Malus x domestica*		
Anna		200	From Israel; adapted to heat of low deserts, Zones 7–9, sf
Arkansas Black		200–800	Tolerates desert heat
Beverly Hills		250–500	Adapted to warm climates, sf
Cripp's Pink/Pink Lady		200–500	Tolerant of Zones 7–9
Dorsett Golden		100–250	From the Bahamas; adapted to hot coastal areas, Zones 7–9, sf
Ein Shemer		100–400	From Israel; good for hotter areas, Zones 7–9, sf
Fuji		200–400	Heat-tolerant, okay in Zones 7–10, sf
Gordon		250–400	Thrives in Sun Belt of California, sf
Pettingill		100–300	Okay in Zones 7–11, sf
Reverend Morgan		400	From Houston, Texas; okay in Zones 7–9
Sundowner		200–300	From Western Australia; does well in South
White Winter Pearmain		400	Best in warm winter areas, sf

COMMON NAME/VARIETY	SCIENTIFIC NAME	CHILL HOUR RANGE	OTHER ADAPTATIONS
Winter Banana		100–400	Okay in Zones 7–9, mild areas of West Coast, sf
Yellow Bellflower		400	Okay in Zones 6–7
Apricots & Apriums	*Prunus armeniaca*		
Blenheim		400	Okay in Zones 5–8, sf
Cot-n-Candy Aprium		250–300	Sf
Flavor Delight Aprium		200–300	Sf
Flora Grenade Pluot		300	Sf
Garden Annie		350–400	Sf
Gold Kist		250–300	Best in warm-winter climes, Zone 8–10, sf
Katy		200–400	Okay in Zones 7–9, sf
Lorna		300–400	Sf
Modesto		300–400	Sf
Cherries, Sweet	*Prunus avium*		
Minnie Royal		200–400	Okay in Zones 7–10
Royal Lee		200–400	Okay in Zones 7–10
Figs	*Ficus carica*		
Black Jack		100	Sf
Black Mission		100	Good in Zones 7–10, sf
Brown Turkey (Vern's)		100	From Texas, okay in Zones 7–10, sf
Celeste		100	Good in cool South, Zones 7–10, sf
Conadria		100	Heat-tolerant, Zones 7–9, sf
Kadota (White)		100	Heat favors ripening, sf
Panache		100	Requires long, warm summer; good in Zones 8–10, sf
Purple Smyrna		100	From Turkey, sf
White Genoa		100	Good in Zones 7–10, sf
Jujubes	*Ziziphus jujube*		
Li		100	Tolerates extreme heat, drought; Zones 6–10, sf
Kiwis	*Actinidia* spp.		
California	*A. chinensis*	250–500	
Issai	*A. arguta*	300–400	
Matua	*A. chinensis*	100–150	
Tewi	*A. chinensis*	100	
Tomuri	*A. chinensis*	100	
Vincent	*A. chinensis*	100	
Nectarines	*Prunus persica*		
Arctic Star		300	Sf
Desert Dawn		150–250	Sf
Desert Delight		100–200	Produces in warm western climes, sf
Nectarina		400	Sf
Panamint		150–300	Excellent in warm-winter climes, sf
Pioneer		300–400	

COMMON NAME/VARIETY	SCIENTIFIC NAME	CHILL HOUR RANGE	OTHER ADAPTATIONS
Snow Queen		250–300	Okay in Zones 7–9, sf
Southern Belle		300	Sf
Sunblaze		250	From Florida
Sunraycer		250	
Sunred		200–300	Sf
Peaches	*Prunus persica*		
August Pride		300	Needs mild winters, sf
Babcock		250–400	Okay in Zones 7–9, sf
Bonanza Miniature		200–250	Sf
Desert Gold		300–400	Sf
Desert Red		200–300	Sf
Donut/Stark Saturn		200–500	Okay in Zones 5–8, sf
Early Grande		200–300	Particularly adapted to South Texas, sf
Eva's Pride		100–200	Sf
Florida Prince		150	Tolerant of desert heat; okay in Zones 8–9, sf
Long Beach		200	Sf
May Pride		175–200	Sf
Mid-Pride		250	Best freestone for Los Angeles & Phoenix to Houston, sf
Red Baron		200–400	Okay in Zones 5–9, sf
Rubichoux		300–400	Sf
Santa Barbara		300	Sf
Saturn		250–300	Sf
Southern Flame		400	Sf
Southern Rose		300	Sf
Tropic Beauty		100–200	Suited to Zone 9, sf
Tropic Prince		150	From Texas, sf
Tropic Snow		150–200	Great in Zone 9
Tropic Sweet		100–200	Hardy in Zone 8–9, sf
Vallegrande		100–200	Sf
Pears, European	*Pyrus communis*		
Comice		200–300	Sf
Floridahome	*Pyrus communis*	300–400	From Florida; okay in Zones 8–10, sf
Hood	*P. communis* X *P. pyrifolia*	100–200	Good for Zones 6–9, sf
Kieffer	*P. communis* X *P. pyrifolia*	200–350	Tolerates hot climates; Zone 6–9, sf
Monterrey	*P. communis* X *P. pyrifolia*	300	Sf
Orient	*P. communis* X *P. pyrifolia*	350	From California, Zones 6–8, sf
Pineapple	*P. communis* X *P. pyrifolia*	150	Good for Deep South; Zones 6–9, sf
Tsu Li	*Pyrus x bretschneideri*	150–300	Good pollinator in warm climates, sf
Ya Li	*Pyrus x bretschneideri*	150-300	Okay in Zones 6–8, sf
Pecans	*Carya illinoinensis*		
Cheyenne		250	

COMMON NAME/VARIETY	SCIENTIFIC NAME	CHILL HOUR RANGE	OTHER ADAPTATIONS
Choctaw		250	Needs long, hot summers, sf
Mahan		250	Best in low deserts, sf
Mohawk		250	Sf
Pawnee		250	Sf
Western Schley		250	Thrives in dry climates, higher elevations, sf
Wichita		250	Sf
Persimmons	*Diospyros kaki*		
Fuyu/Jiro "apple"		200	Okay in Zones 7–10
Hachiya		100–200	Okay in Zones 6–10
Hiyakame		200	
Izu		100	Okay in Zones 6–10
Tanenashi		100–200	Okay in Zones 7–10
Plums	*Prunus* spp.		
Beauty		250–400	Okay in Zones 5–10, sf
Burgundy		150–400	Okay in Zones 5–9, sf
Catalina		300–400	Sf
Green Gage		400	Okay in Zones 5–9, sf
Gulf Gold		250	Sf
Gulf Ruby		250	Good for warm parts of Gulf Coast, sf
Inca		200	
Mariposa		250–400	Good for Zones 7–9, sf
Methley		250	Hardy in Zones 6–9, sf
Santa Rosa		300–500	From California, good in Zones 5–9, sf
Satsuma		300	Good in Zones 5–9, sf
Pomegranates	*Punica granatum*		
Ambrosia		150	Okay in Zones 7–10, sf
Angel Red		150–200	Okay in Zones 5–9, sf
Eversweet		150	Okay in Zones 8–10, sf
Grenada		150–200	Okay in Zones 7–10, sf
Kashmir Blend		150–200	Okay in Zones 7–10, sf
Red Silk		150–200	Okay in Zones 7–10, sf
Sonoran White		150–200	Sf
Wonderful		150–200	Okay in Zones 5–10, sf
Quinces	*Cydonia oblonga*		
Cooke's Jumbo		100	Sf
Orange		300	Good for Zones 6–10, sf
Pineapple		100–300	Okay in Zones 5–9, sf
Smyrna (Izmir)		100–300	From Turkey, sf
Sonoran Membrillo		100–200	From Sonora and Arizona, sf
Walnuts (Persian)	*Juglans regia*		
Pedro		400	Okay in Zones 6–10
Placentia		300	Adapted to mild coastal climates

Planning and Practices

Despite the limitations that changing chill hours may place on your capacity to have a productive orchard with the varieties that you knew in your childhood, many other varieties can be combined into a low-chill fruit guild that will nourish your body and delight your senses. The trick is to plan ahead on a two- to three-decade scale so that the varieties you plant today won't begin a steep decline in productivity due to the reduction in chill hours anticipated over the next quarter century. Of course, the mix of varieties optimally suited to an orchard in the Central Appalachian hollers and coves of North Carolina will still be different from those suited to the Rio Arriba of New Mexico.

If you first decide how many trees you can plant on the basis of the catchment size and volume of rainwater you can store in the soil and in cisterns, you can begin to structure your low-chill orchard from the top down. Here are the steps to creating a multi-strata, low-chill orchard suited to your locality over the next quarter century:

1. Contact your extension agent or state climatologist to determine what your locality's range of chill hours accumulated over the dormant

The Urban Farm Nursery in Phoenix, Arizona, offers low-chill, desert-adapted heirloom fruits for permaculture guild plantings.

Electing Members to Your Low-Chill Fruit and Nut Guild

My friend Michael Phillips has a way with words when he poetically describes what we want to achieve with a fruit guild adapted to any particular foodscape. In his fine book *The Holistic Orchard*,[12] Michael sets out the phyto-socialistic principles that should guide the design of any guild:

To speak of "planting a guild" captures the harmony inherent in a permaculture-inspired layout for

Multi-tiered oasis orchard layers feature a canopy of perennial fruit trees and palms that provides shade for understory vegetables and herbs.

the home orchard . . . A natural community of plants achieves a balance among recycling nutrients, resisting disease, and keeping pests in check while conserving water and attracting beneficial insects and other animals into the fold. Such an ecosystem hums with inner graciousness . . .

But if anyone presumes that Michael's lofty goals cannot be met by a confederation of carefully selected fruits and nuts in the canopy conversing with their understory, microbial, and insect allies, think again. Michael tangibly links what we need in the overstory to the understory to the root zone in a comprehensive and coherent manner. In the overstory, we want what ecologists call "niche partitioning" in the use of light by various trees, as well as wind protection by denser canopies on the upwind side of the orchard. In the understory, we want shade-tolerant herbs, each of which fulfills one or more of the following roles:

- *Nutrient pumpers* (like comfrey), which have deep root systems that mine potassium, calcium, and other minerals, and bring them nearer to the soil surface;
- *Mycorrhizal accumulators* (like ironwood in fredges, lavender, or thyme), which foster diverse and abundant populations of endolithic fungi among the intertwined root systems of trees and herbs;
- *Soil enrichers* (like mesquite), which generate nutrient-rich masses of foliage that can be composted and mulched in place;
- *Weed suppressors* (like peppermint or comfrey), which smother out invasive grasses or nutrient-guzzling annuals with gusto;

season has been in recent past (5 to 10 years), as well as what it's likely to be in another quarter century. Use the latter figure to match up with the chill hour requirements of fruits and nuts you wish to select. If you are in the 100 to 300 range of chill hours, it is likely that

This pomegranate tree (far left) provides partial shade to spinach plants in this multi-tiered guild in Oman.

- *Disease suppressors* (like Papago peas, mints, or chives), which dampen Texas root rot, peach leaf curl, or apple scab, respectively;
- *Insectary plants* (like passion vine or sweet cicely) for the nectar provision, nesting, and roosting of beneficial insects such as syrphid flies or carpenter bees;
- *Rodent ridders* (like red squill, daffodils, and wild chiles) that raise such a stink or leave such a toxic treat that bark-eating voles, prairie dogs, or cotton rats won't mess with the fruit trees;
- *Nectar and pollen producers* (like peppermint, salvia, and wild chokecherries) that keep both resident and migratory pollinators in place during the blooming season of fruit and nut trees; and
- *Insect pest confounders* (like buffalo gourd, desert lavender, or wild oregano), with flowers and foliage that produce so much fragrance, the pests can't find the smell of their target tree or bush.

The game to play with guild design in the new normal is to find understory herbs for the low-chill orchard that are relatively shade- and salt-tolerant, while having a modicum of tolerance to drought and heat. They must still be capable of playing one or more of the above-mentioned roles. While comfrey or chicory might not make it in all hotter, drier climates, mints and lavenders will take the heat and get the work done.

a wide variety of other fruits—dates, olives, guavas, passion fruits, and avocados—may be growable as well.

2. After selecting the particular varieties in the chill hour range suitable for the next century, list them in a table, ranking them by canopy

height, foliage density, blooming period, and tree longevity. Asterisk any varieties that will also require a second pollinator variety for outcrossing, one that overlaps in blooming period with the first. Decide whether you want to mix dwarf, semi-dwarf, and standard sizes (based on rootstock choice) or keep to one size class.

3. Sketch a base map of your available orchard space, and place the larger, denser trees on the upwind side to serve as windbreaks for the others. Scatter other taller, top-canopy trees throughout the orchard. Consider where salt buildup or waterlogging may occur and match more tolerant varieties to these soils or micro-habitats.

4. Next, consider whether you have a top canopy high enough to allow a secondary canopy of shorter trees—pomegranates, pawpaws, jujubes, or dwarfed stone fruits—or of grapevines and berry bushes. Select partially shade-tolerant varieties for this second tier. Also factor in tree spacing and runoff irrigation flows at this stage.

5. Once your one or two tiers of canopies set the structure for your orchard, begin to select understory plants that have an ecological role to play on behalf of the particular fruit and nut trees you have selected. For instance, planting chives in a circle along the drip line of apple trees reputedly lessens their risk of succumbing to apple scab disease. Similarly, planting garlic or shallots around peach trees may reduce the probability that they succumb to peach leaf curl.

6. Once you have placed these specific benefactors to particular fruits or nuts at appropriate places in the understory, begin to add other plant allies: mycorrhizal accumulators, soil enrichers, weed suppressors, pollinator insectaries, rodent repellers, and so on. Make sure your selections are tolerant of partial shade, and, depending upon the soil, tolerant of salts and occasional waterlogging. Because they will spend most of their lives under the tree canopies they need not be as drought- and heat-tolerant as open meadow plants, but they should not be water guzzlers that compete with the trees for moisture.

7. Use your map to begin to build a guild that will change slightly in composition over the next century—particularly in the understory—but will offer the same resilient structure and function. Gauge water delivery rates and volumes to the ages and canopy densities of the trees.

8. Use both your own taste buds and a handheld **refractometer** to measure soluble solids (Brix levels) in the fruits to guide you in soil nutrient management, thereby increasing the nutrient density of the food you harvest. Taste, see, and enjoy!

Chapter Eight

When Terraces Are Edged

with Succulents and Herbaceous Perennials

❀

Warm-Up

Standing at the rim of a high desert plateau in Oman named Jabal al-Akhdar—an ancestral home of my kin in the Banu Nabhani tribe, from which I am descended—I counted 28 terraces cascading down the slopes into a deep canyon. I stood there amazed. Earlier in my life, I had seen spectacular terraced agricultural landscapes in Mexico, Ecuador, and Peru, where each terrace edge is typically planted with prickly pear cactus or century plants like agaves, with their rosettes of sword-shaped leaves. But this was an altogether different desert with what I presumed to be a different suite of crops, and so I was eager to see which of their 100-some crop species my Arab relatives chose to place on their terraces.[1]

What I could not see from the rim of the plateau—but what surprised me 20 minutes later when I had descended to the middle rung of terraces—was that the edges of these terraces on the Arabian Peninsula were often adorned with prickly pears *and* century plants! Although desert farmers in the Middle East have only had access to these succulents from the Americas for less than 500 years, they had so well incorporated them into their terracing designs that one might assume they had been there forever.

It seems that in the hot, dry climes of most deserts, semi-deserts, and subtropical regions, succulent perennials and dry-stone terraces go together like beans and corn bread or hummus and pita. And yet in the drier reaches of the United States, which were, during late prehistoric times, covered by hundreds of thousands of acres of terraced agave fields, it seems that Americans have succumbed to some sort of perennial agricultural amnesia. Most farmers in the Desert Southwest today don't even consider prickly pears and century plants to be viable food crops; worse

These ancient terraces on the otherwise dry Jabal al-Akhdar plateau in Oman have supported food production for centuries, if not millennia.

yet, they hardly ever build stone terraces on the slopes of their farms or orchards, or regard such habitats as arable lands.

While some of the best wine grapes and olives in the world are grown on stony, semi-arid terraced slopes with soils much like those in the American Southwest, North American farmers for the most part stick to cultivating floodplains and gently rolling plains where massive tractors and combines can move around unhampered by the land's constraints. There, our ancestors have broken up the sod, eliminated the perennial cover of legumes, grasses, and wildflowers, and let some of the best soil in the world simply blow away. At the same time, a century of poorly managed grazing on slopes and plateaus considered worthless for crop production has aggravated erosion, creating head cuts and gullies on land that would be prized for terraced gardens and orchards in Morocco, Spain, Italy, Greece, Turkey, Lebanon, or Syria.

We have neglected caring for the soil at our own peril. As Edward Hyams suggested more than half a century ago in *Soil and Civilization*, how a society treats or mistreats its soil presages its destiny, economic trajectory, or collapse.[2] The *National Lampoon* once argued that the United States should change its currency to the silt eroded away from our productive lands, because that has become the product that we are most apt to produce and deliver to those who live downwind or downstream from us!

Perhaps that is why the consequences of how some American farmers responded to the 2012 drought will be felt for many years to come. Before that year was over, soil scientists in the Midwest expressed concern that because farmers had opted to harvest entire corn plants for silage

Agaves and terraces go hand in hand to reduce soil erosion from slopes.

The Ancient Arab Terraces of Jabal al-Akhdar, Oman

While most Americans assume that the majority of Arabs were, until recently, nomadic herders, more sedentary *al-hadr* tribes of the Arabian Peninsula have been farming the oases in the valleys, canyons, and slopes of the mountain country of Oman and Yemen for millennia. It is hypothesized that they are the peoples who first devised desert irrigation agriculture. Since their origins on the southern peninsula, their techniques for constructing and even their term for irrigation canals, *as-saqiya*, have traveled the world, and have been transformed into the *acequia* systems of Latin America and the Desert Southwest.

On the flanks of Jabal al-Akhdar—the high desert plateau that has served as the ancestral home for many of my distant Nabhan kin—I saw remarkably intact terrace systems based on water harvesting and containment strategies that have stood the test of time, while modern agricultural developments nearby that were based on pumping highly alkaline groundwater have salinized the soils and collapsed within 20 years of being constructed.[3]

While not all the ancient terraces below the villages of Wadi al'Ayn and Sheraga have been maintained, it is clear that recent efforts have gone into building stronger retaining walls for the terraces still in use. Water is collected off the barren bedrock stretches close to the edge of the plateau, and funneled into *falaj* (*qanat*-like) tunnels that store desert storm runoff; this water is then released into irrigation canals that cascade down canyon slopes, meandering from one terrace to the next.

German agro-ecologists have documented no less than 107 crop species in the terrace gardens of Jabal al-Akhdar, and these may be represented by hundreds of place-based varieties.[4] Depending on the width of the terrace, annual field crops such as onions, garlic, spinach, and radishes may be grown in furrow-irrigated patches, or fruit trees such as pistachios and pomegranates, apricots and almonds, limes and papayas may be grown in flood-irrigated basins. Dense patches of the rose of Damascus are cultivated to produce rose water for "export" to Arab villages in the hotter, drier lowlands. The rims of some of the terraces are edged with two hardy cacti of American origin, *Opuntia ficus-indica*, the cultivated thornless prickly pear that Luther Burbank made famous, and *O. humifusa*, a thorny prickly pear that was introduced to Egypt around 1920. The fiber-producing *Agave americana* can also be seen edging the lips of terrace walls in this region of northern Oman.

Sadly, global climate change is now knocking out some of traditional winter-chill-requiring fruit and nut varieties grown on Jabal al-Akhdar terraces, because the number of chill hours has decreased in the mountain oasis microclimates by as much as 275 hours between 1983 and 2012. Except in the coolest oasis village on the plateau, crops such as walnuts, peaches, and apricots are no longer receiving sufficient chill hours for flowering and fruiting, and even pomegranates are not expected to always receive sufficient winter chill over the next few decades.[5] Omani farmers have begun to shift production of chill-requiring varieties to cooler terraces or even to the coolest oasis villages, but the varieties requiring the greatest chill hours will likely be lost from most of these agrarian landscapes.

rather than ears for grain alone, they were potentially depleting their soil bases. By removing more biomass from fields rather than incorporating it on or into the soil during the 2012 growing season, farmers had left a considerable amount of agricultural land bare and susceptible to erosion. Although the effects of soil erosion on productivity can vary between years due to rainfall differences, soil erosion over time will reduce a land's yield potential. Research has shown that corn yields in eroded land can be between ten to thirty percent lower when compared to un-eroded soil.[6]

Of course, even before the massive drought of 2011–12 hit them, many American farmers were not necessarily covering and caring for their soils as diligently as they should have. Despite eight decades of federal agencies and private landowners investing in soil erosion control, between one and two billion short (US) tons of topsoil from croplands is annually lost from agricultural use, and that volume makes up about a quarter to a half of all the soil lost in the entire country.[7] Every two decades, the average field or orchard in America loses a third of an inch of topsoil, or about 50 short tons of soil moisture-holding capacity per acre.

The worst erosion—as you might expect—is on steeper, unterraced slopes, where sheet and rill rates may soar as high as 23 tons of topsoil loss per acre in a single year, and average 15 tons per year over the last two decades.[8] While the loss of topsoil per year has been gradually declining since the Dust Bowl era in the 1930s, it may well be because there is relatively less to lose in areas that have already been highly eroded. Nevertheless, nearly one-fifth (19.3 percent in 1997) of all croplands in the United States remain affected by moderate, severe, or extreme water erosion, with non-irrigated cultivated lands being more dramatically affected.[9] Add the damage done by wind erosion to that done by the sheet and rill water erosion noted above, and more than 100 million acres of agricultural lands in America are still losing topsoil at significant rates. Worldwide, it is estimated that as much as 40 percent of all agricultural lands today are severely eroded.

Fortunately, there is renewed interest in examining which land stewardship practices foster soil resilience. In this case, *resilience* is a code word for the capacity of living (microbe-rich) soils to absorb, resist, or recover from climatic changes or other natural or human perturbations. A resilient soil retains most of its mass, and quickly recovers its nutrient profile, water-holding capacity, and biodiversity after a natural disruption or human insult.

Of all the strategies traditional farmers around the world have implemented to reduce erosion, retain moisture, and sustain soil resilience, perhaps the terracing of sloping lands is the oldest and most widely practiced.[10] In general, agricultural terraces on the slopes of ridges, hills, and mountains reduce the steepness of slopes, intercept runoff, slow the force of running water, which then encourages infiltration and the deposition of nutrients. Well-constructed and diligently maintained terraces can reduce runoff by 25 percent and minimize soil loss to one-twentieth of their

A Tarahumara Indian stone wall terrace holds a maize field's soil in place on this slope in the Sierra Madre Occidental in Mexico. PHOTO BY LAURA SMITH MONTI.

former rates on erosion-prone slopes, keeping more moisture and nutrients available to the crops planted upon them.[11]

And yet terracing in and of itself is not a panacea. Not all types of terraces are created equal in their capacity to buffer croplands from extreme climatic events. Even the best-constructed terraces are seldom able to permanently reverse erosion.[12] The most effective terracing systems almost always need to have perennial plants such as prickly pears and century plants deeply rooted on their outer edges. Furthermore, most historically constructed agricultural terraces are no longer sufficiently maintained today, given the labor costs of continuous repair and care of such soil conservation structures. It behooves each of us to go out and learn what kinds of terraces and perennial plants have worked best together for the longest periods of time, and the pledge to maintain in the particular landscapes where we live and grow food.

Parable

In the late 1980s, I happened to be present and humbled when a remarkable archaeological discovery was made in Arizona—the presence of living century plant populations on terraces and rock alignments constructed

500 to 800 years earlier. This marked the first time that scientists confirmed that the very same crop variety (or genetic lineage) has survived in an unbroken chain since prehistoric farmers had first planted it on terraced hillsides centuries before us. What's more, the volcanic cobbles laid down in rock alignments and low terraces by desert dwellers 25 to 40 generations before us still held soil and moisture in place long after they had been abandoned.

Like most so-called discoveries, ours was a fortuitous find; many other archaeologists and botanists with just as much (or more) expertise as our own had tramped along the same slopes without ever glimpsing the remnants of ancient crops surviving on those terraces. Some of my own archaeology professors had found an extensive array of terraces, contoured rock alignments, check dams, rock piles, and waffle gardens in the vicinity of Table Mesa some two decades earlier, but had no clue from their excavations and their pollen samples about what had been grown there.[13] At the same time, a gifted horticulturist and field botanist, Rick DeLamater, had mapped dozens of sites near the mesa where a rare century plant grew, but he had assumed they were merely wild populations. Having once surveyed plants in the area with archaeologists, I was curious to learn whether any of the century plant known as *Agave murpheyi* could still be found on the prehistoric terraces in many spots surrounding Table Mesa. Indeed, it has now been rediscovered in several watersheds near Table Mesa.

Agaves persisting on historic terraces in the semi-arid canyons of southern Arizona.

The Story Behind the Terrace at Perry Mesa

Recently a team of ethnobotanists, archaeologists, and soil scientists looked at the soil accumulated behind terraces and rock alignments at Perry Mesa, another prehistoric settlement not far from Table Mesa. They were eager to determine whether or not the ancient agricultural systems there had been truly sustainable during their course of operation. Up until the early 14th century, water harvesting for food production was scattered across 250 acres of hillside terraces and floodplain catchments at Perry Mesa, and likely fed 2,500 to 3,500 indigenous residents at any one time.

The team found that some of the rich sandy loams behind the terraces had been created by blending clays or sands on site with other "imported" materials that had probably been carried up from the floodplains below the terraces.[14] The imported materials were much like those found in biochar-enriched soils in the Amazon Basin:

charcoal, ash, ceramic fragments from broken pottery, and, in some cases, animal bones. This composted fill was roughly a foot deep behind some multi-coursed terraces built on steep (16 percent) slopes where such a depth of soil does not necessarily accumulate naturally. Other terraces had clay loams behind them that retained more soil moisture during dry years, but they had become somewhat compacted and more alkaline over time due to either agricultural activity or foot traffic.

Nevertheless, the overall impression gained by the team from Arizona State University was that the terrace soils in the vicinity of Perry Mesa and Table Mesa were significantly different from untended desert soils. The terrace soils continued to hold moisture, retained agriculturally suitable levels of buffered alkalinity, and sustained nutrient levels still functional for food production centuries after they were first prepared.

And so I invited a few of my archaeological mentors—Vorsila Bohrer, Paul Fish, and Suzanne Fish—to come into the field with a number of botanically trained field scientists, including Rick DeLamater, Wendy Hodgson, Donna House, and myself. Within 50 yards of leaving our cars on the sole rocky road that climbed up the mesa, we spotted a few century plants on the lower slopes or volcanic *bajada* that flanked the mesa. As we gathered around the plants to see exactly where they were situated, I pointed out the prehistorically constructed rock alignment that followed the contour for dozens of yards across the slope. Another of us realized that this clonal population of century plants was still situated on the very same terrace created by a rock alignment built hundreds of years prior to our own arrival there!

And then—to our astonishment—one of the archaeologists reached down between the century plants and found what prehistoric tool experts call an agave knife. It was a thin piece of carefully shaped stone that had a curvilinear blade with obvious serrations along its cutting edge. Our jaws dropped. Here were the very same crop plants, terraces, rock alignments, and harvesting tools that had fed a prehistoric farming family a half a millennium ago![15] What's more, many of the terraces still retained enough fertile soil and moisture that both agaves and a fishhook cactus with delicious strawberry-like fruits continued to produce abundant food crops long after the last cultivator had departed from this ancient agricultural landscape five to six centuries ago.

Wendy Hodgson and her colleagues have since demonstrated that *Agave murpheyi* populations have been genetically diverse enough to withstand all sorts of climatic shifts.[16] And yet they were clonally propagated much the same way that bunching and walking onions, as well as maternity plants, are produced: by vegetative offshoots at the base of the mother plant, or by bulbils (miniature versions of the mother) on the flowering stalk. When the bulbils or plantlets fall to the ground, someone must disturb the soil, or nuzzle them up against dew-collecting cobblestones, or else they will wither and die. And so many of the prehistorically cultivated agaves were intentionally planted in loosened soil and then half buried in rock piles, where dew and light rains would collect on the bottom sides of the cobblestones in the pile. This enhances the moisture available to each plant.

Principles and Premises

If we take seriously the need to conserve soil as well as water *in place*, then how we grow food in the headwaters of a watershed may be as important as what we do in the valleys below. For that reason, any orchards, fields, or gardens perched on the high slopes of plateaus, mesas, ridges, hills, or mountainsides deserve to be terraced *and* edged with herbaceous perennials with extensive root systems that can hold soil structure in place.

That said, we should remember that there are many different kinds of terraces that can be constructed for different degrees of steepness on slopes, for different soils and parent materials (rock types), and for different growth forms of crops. Some kinds of terraces are costly in terms of the investment in labor and materials required for their construction and maintenance, while others are not. Ted Sheng, a watershed scientist who has worked extensively with traditional farmers in Asia and Latin America, has devised a simple but rather complete typology of terraces for placement on steep to gentle slopes.[17] From my own informal field surveys of terraces in Latin America, Central Asia, and the Middle East, I have modified and amplified Sheng's menu of terrace types to some degree:

- Long, continuous **bench terraces**, which are **reverse-sloped** (angled to retain runoff) for upland crops;
- Shorter, intermittent bench terraces for upland crops, which can be constructed as modules and then coalesced into more continuous terraces over several years;
- Hillside ditches or rock-lined canals for excess runoff from areas planted with short-lived perennial crops, constructed as narrowly terraced drainage ways;

Check dams not only slow erosion, but can revive intermittent or perennial stream flow on farms or ranches, as Joe Quiroga has done over the past 15 years on the Diamond C Ranch near Elgin, Arizona. PHOTO BY GARY PAUL NABHAN AND CALEB WEAVER.

Parameters for Building a Bench Terrace

The most common types of long-running or intermittent terraces in slopes are variations on the **Nichols bench terrace** that has been constructed over the millennia on several continents. Given its broad adaptability, it is worthwhile to recap the design principles that Ted Sheng has articulated for bench terraces. Sheng has developed simple computer programs to guide terrace builders and farmers in the quantitative calculations required to design durable terraces after studying and constructing them himself in several arid and semi-arid countries.[18] The following parameters are gleaned from Sheng and many others, but can be examined in more depth in Sheng's manuals and articles, available online:

- The width of the bench (the flat surface of the terrace) should be determined by the farmer according to amount of runoff and soil volume needed for the particular crops to be planted, the kind of tools (or machinery) to be used for tillage and harvesting, and ease of access. Sheng has observed that, for manual construction and hand cultivation, terraces can be 7 to 15 feet wide, but if the work is mechanized, 10- to 24-foot-wide terraces are more manageable.
- If the steepness of the slope is in the 12.5 to 50 percent range, the bench terrace can likely be constructed with hand tools (picks, pulaskis, and shovels) by cutting into the natural slope just below the first flat area, and using the displaced earth to build up the slope below the excavation to form what is known as a Nichols bench terrace.
- All of the fill required can usually come from the site itself, unless the landscape is extremely rocky or underlain with layers of caliche (cement-like calcium carbonate). However, adding biochar, compost, or other soil amendments from nearby is usually advisable.
- The terrace bank, riser, or retaining wall should largely follow the contour of the slope, and that contour can be established by using a spirit level or bunyip, and then flagging it to establish the contour line of the terrace edge.
- The base of the retaining wall, riser, or bank needs to be strong, and protected by an armor of rock or brick to keep it from being undercut during violent rainstorms.
- Perennial vegetative cover is useful in helping protect the terrace lip from being eroded by forceful runoff, and to protect the base of the retaining wall, riser, or bank from being undercut. Deep-rooted vines and scandent (sprawling) shrubs such as spiny capers and desert-adapted succulents such as prickly pears and century plants not only stabilize the terrace edge, but can serve as barriers to keep goats, sheep, or cattle from entering the terrace plantings.

Agaves on reconstructed hillside terraces at Trincheras, Sonora—a Mexican monument to water harvesting for food production.

- Convertible terraces behind check dams and gabions situated in sloping watercourses, which may build soil behind over time, so that they can be used later for planting;
- Orchard terraces, which are typically wide enough to underlie the entire canopy of a fruit or nut tree;
- Hexagonal **balcony** and/or **window box terraces**, which support small orchard clusters perched on very steep slopes;
- Bowl-like basin terraces that hold single trees or dense clusters of herbaceous perennials; and
- Naturally formed curvilinear terraces that are slightly modified to make them more suitable for patches of upland annual crops.

No matter what the shape or size of terrace, the kinds of crops to be planted on it should match the available area of arable soil, the kind of soil (with respect to its capacity for retention of runoff from upslope), and the ease of access for carrying out the harvest. The planted area can comprise two components: the tree crop or herbaceous crop planted in the core area, and the herbaceous perennial crop planted on the outer lip or rim of the terrace.

Planning and Practice:

Contoured Bench Terraces for Mescal Century Plants and Prickly Pear Cacti in a Permaculture Setting

As noted in the first part of this book, century plants and prickly pears are among the most water-use-efficient crop plants in the world, producing two to four times the biomass for the same amount of water as most conventional annual crops. They are widely grown on terraces for food, fiber, and beverages in most semi-arid, arid, and dry subtropical parts of the world, *except* in the present-day United States. The following bench terrace design builds on the premises, principles, and parameters that have been presented in the previous section of this text, and is adaptable to many geomorphic contexts. However, the particular conditions that I faced when constructing my own terraces with permaculture designer Caleb Weaver are also presented in the sidebar for those of you who need tangible numbers from one specific design in order to envision your own.

Because we will be considering the placement of only a single row of succulents on each cobblestone-armored terrace in this practicum, we hardly need more than 78 inches (2 meters) of flat surface on each terrace to allow for the growth of the succulent plant, and to provide enough room for a harvester or pruner so that he or she will not get "stuck" by the plant's spines. In this case, we will assume that the average slope of the ridgeside where the terraces are to be constructed is about 35 percent, or a little more than 20 degrees; you can measure your own slope with a clinometer or handheld level if you wish. Given these parameters, let us see what kinds of information can be generated to guide our design of the bench terraces.

One of hundreds of prehistoric hillside terraces at Trincheras, in the heart of the Sonoran Desert.

Bringing Hillside Terrace Cultivation of Cacti and Century Plants Back to Southern Arizona

For the last three years, I've been developing ridgeside terraces in Patagonia, Arizona, on which to evaluate the climatic tolerances of prickly pears (*Opuntia* and *Nopalea* spp.), as well as mescal and sotol (*Agave* and *Dasylirion* spp.). Because some of these plants are heat-tolerant but freeze-sensitive, while others are cold-hardy but heat-intolerant, my evaluations at 4,000 feet in elevation may aid growers who live at elevations that are either higher or lower than where I live and farm. I grow a dozen varieties of prickly pears belonging to four species (*Opuntia engelmanii*, *O. ficus-indica*, *O. streptacantha*, and *Nopalea cochenillifera*), one species of sotol or desert spoon (*Dasylirion wheeleri*), as well as several species of agave (*Agave gentryi*, *A. huachucensis*, *A. ovatifolia*, *A. palmeri*, *A parrasana*, and *A. parryi*). Others, such as bacanora and tequila agaves that were gifted to me, died in a catastrophic freeze that occurred in the Southwest in February 2011.

All of these succulent perennials are grown as part of my **Zone 4** permaculture plantings—the second-to-last, semi-wild domain of minimally managed food species—with no supplemental irrigation after initial establishment. On a 90-foot-long southwest-facing ridgeline, we have constructed eight cobblestone-line terraces that cascade down over a 40- to 50-foot slope, with the terrace intervals ranging from 4.5 to 6 feet. The width of these intervals depends upon the existing topography, especially the steepness of the slope. The vertical height of the terrace banks varies from 2 to 3 feet, with two to five courses of cobblestones armoring them. All agaves are transplanted after placing biochar, mycorrhizae, and other effective microbes in their holes. Although the prickly pear fruits are already being harvested for fresh eating and for "nectar" or syrup, the century plants and sotol will take several more years to mature. However, it is already evident that the supplemental runoff they receive on the terraces is speeding up their growth, so that some may mature and flower in a matter of five to six more years. Meantime, they are holding in place the terrace soils on a slope that had been previously vulnerable to erosion.

1. Walking the ridge side with a clinometer or spirit level in hand, flag the areas that most deviate from the average 35 percent (20-degree) slope gradient, and then flag those places where the contours may need to be placed at somewhat longer or shorter intervals.

2. Find a flat place on or near the top of the ridge that averages 6 feet or more in width. With a bunyip, or with a spirit level and horizontal pole or board, mark the top contour for where you would like the terraces to begin.

3. To proceed toward discerning the vertical interval between terraces, you will need to factor the slope of the retaining wall or riser. Assume a 0.75 slope for a hand-excavated bank or stable clay and stone, or a 0.5 slope for dry-stone retaining walls. Machine-built, reinforced concrete or brick risers can achieve a slope of 1.0 in many cases.

Agave, sotol, and prickly pear cactus can be planted on hillside terraces and produce yields without supplemental irrigation.
DRAWING BY PAUL MIROCHA.

4. To obtain the ideal vertical interval (VI) between terraces, use the following equation offered by Sheng:

$$VI = (S \times Wb)/100 - S \times U)$$

Where S is the percent slope (35%), Wb is the width of the bench or flat part of the terrace (2 meters), U is the slope of the wall (1.0 for concrete, 0.75 for hand-excavated clay and stone, or 0.5 by dry-stone retaining walls). Here we will use the 0.75 slope. The formula determines that the theoretical vertical interval for the retaining wall will be 37 inches or 0.95 meter, as shown below:

$$.955 = (35 \times 2 \text{ m})/100 - (35 \times 0.75)$$

Intercropping perennial artichokes with annual shade-giving sunflowers beats the heat and holds soil in place.

Simple but elegant terrace walls of willow osiers can be held in place by rebar posts in order to retain soil and moisture.

5. Given those dimensions, the width of the riser or bank should be no less than 16.5 inches or 0.43 m. If the linear length of the entire terrace on contour is to be about 3,000 feet (900 meters), the volume of soil required from cutting to fill an even-planed terrace would be nearly 360 cubic yards (260 cubic meters).

6. Because there may be considerable rock in the fill, it is wise to sort any cobblestone-sized pieces out and use them to armor the base and top of your clay-and-stone retaining wall. Replace the volume of stones that you remove from the fill with an equal volume of compost, mulch, or local flood-washed detritus.

7. If you choose instead to build a dry-stone retaining wall, then dig away at the base of your vertical cut in the clay and stone bank until you have a flat surface 18 to 24 inches wide, and begin to place flat-sided stones against the base of the bank, with their more even sides face-out. Place the rougher edges of the cobblestones against the bank, tilting them against the cut of the slope. After one layer is set, fill in clay as a moist batter in between and atop the cobblestones. Fill in any spaces between the stones and the cut of the bank, and then begin the next course. Ultimately, the armored dry stone wall will need 1 foot of batter for every 3 feet (0.95 meter) of wall height.[19]

8. Cap the wall, riser, or bank with the longest cobblestones you can find, so that they sit a few inches above the soil grade of the terrace. Place filter fabric immediately behind the capstones and cover it with coarse gravel to prevent runoff from washing out the soil between the capstones.

9. Select a suite of succulent or herbaceous perennials to plant immediately inside the terrace from the capstones. Depending upon their ultimate size, place transplants at least 9 inches back from the capstone edge, and at least 1.5 to 2 yards apart from one another. Select perennial species with particular varieties that are adapted to your current and future climate (see table 8-1), and plant no less than three species and six varieties along the terrace edges. Add biochar, broken pottery, compost, and mulch to their holes when transplanting them out.

Terraces and the perennial plants require regular maintenance, but if well placed to begin with they may last for decades. Perhaps our true test of sustainability is when the perennial succulents on our terraces—and the terrace walls themselves—come to outlive each of us!

TABLE 8-1 Succulent and Herbaceous Edible Perennials for Planting on the Lips of Bench Terraces

SCIENTIFIC NAME	COMMON NAME(S)	RECOMMENDED VARIETY	ORIGINS, USES, & ADAPTATIONS
Agave murpheyi	Hohokam agave	Variegated, bulbil forming	From Sonoran Desert lowlands in Arizona and Sonora; most plant bulbils are planted in loose soil, but are very drought- and heat-tolerant once established
Agave palmeri	Palmer's agave, lechuguilla	Some exist, but they are not yet formally named	From low and mid-elevations in the Southwest and Mexico, where frost-hardiness is essential
Agave parrasana	Cabbage head agave	Fireball	From Mexico, a high-elevation variegated mescal that flowers in late spring
		Meat Claw	Cold-hardy to 10°F (-12°C), and heat-tolerant to over 100°F (38°C)
Agave parryi	Parry agave, mescal de los Apaches	Sunspot	From New Mexico, cold-hardy to -10°F (-23°C) and heat-tolerant as well
Agave potatorum	Butterfly agave, Papalometl	Kichiokan/Snowfall	From high elevations in Mexico, sun-tolerant
Allium ampeloprasum	Leek	Elephant garlic	Heat-tolerant, but needs winter protection from cold
		Inegol	Slender and tender variety from Turkey
		Kurrat	Rare heirloom from Egypt
		Tarée Irani	Rare salad leek from Iran
Allium cepa var. *aggregatum*	Bunching onion, shallot	I'itoi's Spanish Shallot	From Arizona, heat-tolerant, year-round producer in the desert
		Pacific Pearl	Day-length-neutral; 50 days to maturity
		Joe's Shallot	Louisiana heirloom
Allium cepa var. *proliferum*	Topset or walking onion	Catawissa	Extremely hardy, but somewhat intolerant of extreme heat
		Egyptian, Moritz	Missouri heirloom
		McCullar's Topset, Pran	Hardy, derived from heirloom from Kashmir and Pakistan
Allium sativum	Garlic	Ajo Morado, Ajo Rojo	Spanish softneck type, originally from Spain; now common in Sonora and Nevada
		Dushanbe	From Central Asian deserts, matures very early
		Israeli	Desert-adapted, large bulbs
Allium longicuspis		Persian Star, Samarkand	From Central Asia
		Sonoran	Turban-type garlic from northwestern Mexico deserts; heat-tolerant and very early maturing
Arctium lappa	Burdock, hobo	Takinogowa Long	From Japan; hardy, 105 days to maturity
		Wantanabe Early	From Japan; biennial, 150 days to maturity
Armoracia rusticana	Horseradish	Bohemian, Maliner Kren	Early blooming, hardy

SCIENTIFIC NAME	COMMON NAME(S)	RECOMMENDED VARIETY	ORIGINS, USES, & ADAPTATIONS
		Variegated	Variegated leaves by second year, less aggressive
Asparagus officinale	Asparagus	Apollo	Dioecious (male and female plants)
		Atlas	Dioecious, drought-tolerant
		Barr's Mammoth	Tolerant of many kinds of weather & rust
		Grande	Dioecious, drought-tolerant
		Jersey Giant	All-male selection, with good yield stability
Basella alba	Malabar spinach, poi sag	Red strain, Rubra	Heat loving, 70–100 days to maturity
Capparis spinosa	Caper	Aculeata	From Mediterranean and North Africa
		Dolce di Filicudi e Alicudi	From the Aeolian Archipelago
		Nocellana	From Italy; spineless, with globose buds, mustard-green color, and strong aroma
		Nuciddara, Nucidda	From Italy
		Spinosa Comune	An Italian form with stipular spines
		Testa di Lucertola	From Italy
		Tondino	From the island of Pantelleria
Capsicum annuum var. glabriusculum	Chiltepin	McMahan's Texas Bird Pepper	Native to southwest Texas; heat-tolerant
		Sonoran	From northern Mexico, heat- and drought-tolerant, frost-sensitive
		Texas/Big Bend	From Texas and Coahila, heat-, drought-, and alkalinity-tolerant
		Tumacacori	From Arizona, the northernmost wild chile; heat- and drought-tolerant, somewhat tolerant of cold winters
Capsicum frutescens	Tabasco		Of unknown origin in Mexico, heat-tolerant
Cichorium intybus	Chicory	Baxter's Special	From Texas; heat-adapted San Pasquale
		Early Treviso	From Italy; 80 days to maturity
		Magdeburg	From Sicily; coffee-substitute chicory
Cnidoscolus acontiflius	Chaiya, Mayan tree spinach	Chayamansa	From Mexico; the most common, heat-tolerant chaya spinach; green without spines
		Estrella	From Mexico; a heat-tolerant green with star-shaped leaf lobes
		Picuda	From Mexico; a heat-tolerant green
		Redonda	From Mexico; a heat-tolerant green with roundish leaves
Cucurbita ficifolia	Chilacayote, figleaf gourd	Alcayota	From Mexico and Guatemala; a high-elevation-adapted, watermelon-like squash with good keeping qualities

SCIENTIFIC NAME	COMMON NAME(S)	RECOMMENDED VARIETY	ORIGINS, USES, & ADAPTATIONS
Cynara cardunculus	Cardoon	Cardon d'Algiers	From Algeria; a variety with huge spiny leaves
Cynara scolymus	Globe artichoke	Kiss of Burgundy	From France; a heat-, drought-, and cold-tolerant rosette
		Purple Sicilian	From Sicily; a good yielder under hot conditions
Hibiscus acetosella var. *sabdariffa*	Roselle	Thai Red	From Thailand; a heat-tolerant beverage plant
Helianthus tuberosus	Jerusalem artichoke, sunchoke	Fuseau	From France; selected for early maturation
		Jack's Copperclad	From the US; a good-yielding, hardy heirloom
		Stampede	From the US; high-yielding, extra-early-maturing, winter-hardy
Lablab purpureus	Hyacinth bean	Lignosa, purple-flowered, Murasakiirohana Fujimame	From Japan, a bean requiring 100 days to maturity; grown by Thomas Jefferson in Virginia as early as 1802
Momordica charantia	Bitter melon	India Long White	From India, China, and Pakistan; adapted to warm climates
Nopalea cochenillifera	False prickly pear	Nopalea grande	From Mexico; drought-tolerant, with edible fruit
Opuntia ficus-indica	Nopal, prickly pear	Amarilla montesa	From Mexico; a long, yellow-fleshed fruit with few seeds
		Amarilla redonda	From Mexico; a round, yellow-fleshed fruit with few seeds
		Burbank Spineless	Really from Mexico, but linked to Luther Burbank in California
		Burrona	From Mexico; a large, wide, late-maturing fruit with light green flesh
		Cardona	A disease-resistant plant from Mexico, with late-maturing, purplish-pink-fleshed fruit
		Charola	From Mexico; light green-fleshed fruit with a gorgeous pink blush
		Cristalina	A high-yielding plant from Mexico, with big, heavy light green-fleshed fruit
		Esmeralda	From Mexico, with light-green-fleshed fruit
		Fafayuco	A late-maturing fruit from Mexico with sweet, white to light green flesh
		Mexican Sweet	From the US, hardy to 26°F (-3°C)
		Naranjona	In Mexico, the most widely grown cactus with orange-yellow flesh
		Queen (Reyna)	The most commercially important cactus variety in Mexico, with early-maturing light green to white flesh
		Roja pelona	The biggest and longest *tuna* fruit in Mexico, with deep red fleshy fruits that mature early

SCIENTIFIC NAME	COMMON NAME(S)	RECOMMENDED VARIETY	ORIGINS, USES, & ADAPTATIONS
Pachyrrhizus erosus	Jicama, yam bean	Jicama de Leche	From Mexico; drought-tolerant
Phaseolus coccineus	Ayocote, runner bean	Aztec White	Quick maturing, relatively heat-tolerant
Polymnia edulis	Bolivian sunroot, yacon		From South America; a sun-loving, jicama-like root crop with edible leaves
Porophyllum coloratum	Papaloquelite	Papalo	A delicious spinach-like green
Rumex scutatus	French sorrel	Buckler's, Silver Shield	From France; a tart-tasting, heat-reflecting, cold-tolerant sorrel
Sechium edule	Alligator pear, chayote, mirliton	Louisiana Mirliton	From the Caribbean and Gulf Coast, a heat-tolerant chayote squash
Scorzonera hispanica	Black salsify	Belstar Super	From Europe; matures in 80 days; a black-skinned, white-fleshed root vegetable
Taraxicum officinale	Dandelion	Ameliore	From Southern Europe; a heat-tolerant, chicory-like green
Yucca baccata	Banana yucca	None yet named	From the US and Mexico; a heat- and drought-tolerant succulent with edible fruit

- Chapter Nine -

Getting Out of the Drought

Intercropping Quick-Maturing Vegetables and Grains in Place-Based Polycultures

Warm-Up

The droughts of 2012 desiccated 71 percent of the annual crops of grains and vegetables in the United States, but other countries such as Canada, Kazakhstan, Mexico, and Russia were just as hard hit. Fortunately, unlike some of those countries, the United States had a formal policy in place that offered crop insurance to the likes of corn and soybean farmers in the Midwest, almost half of whom suffered poor yields or outright crop failures in 2012. Although the drought's effects on annual crop losses were perhaps comparable in severity to those suffered in the mid-1950s and the 1930s, the amount of crop insurance doled out in the United States in one year has reached a record high $20 billion. The cost of foods such as corn and wheat hit new peaks in global markets as well, and an increasing number of poor and hungry people found that even the prices of staple cereals were suddenly put beyond their reach.

And yet for all the crop insurance payments offered and all the federal and local investments made in food banks, soup kitchens, and commodity food deliveries to buffer Americans from hunger in the face of climatic disasters, one fundamental form of crop insurance and food security was neglected or ignored altogether: agricultural biodiversity. Let me broadly define this term for you as I did in an earlier book, *Where Our Food Comes From*,[1] because it may remind you how much any successful strategy for dealing with climate change will be tied to wisely using these resources, as the surest form of "crop insurance":

Bean diversity includes many early-maturing, short-cycle varieties that are adaptable to climate change.

Agricultural biodiversity is embedded in every bite of food we eat, and in every field, orchard, garden, ranch and fish pond that provide us with sustenance and with natural values not fully recognized. It includes the cornucopia of crop seeds and livestock breeds that have been largely domesticated by indigenous stewards to meet their nutritional and cultural needs, as well as the many wild species that interact with them in food-producing habitats. Such domesticated resources cannot be divorced from their caretakers. These caretakers have also cultivated traditional knowledge about how to grow and process foods; such local and indigenous knowledge—just like the seeds it has shaped—is the legacy of countless generations of farming, herding and gardening cultures.

While previous chapters may have made you marvel at how many varieties of fruit and nut trees, as well as herbaceous and succulent perennials, you can potentially employ in adapting our food production to hotter and drier conditions, the simple fact is that *most of the **agro-biodiversity** available to farmers remains in the form of annual crops*, including cereals, leafy vegetables, roots, and shoots. However perennialized we wish our agriculture to become, most of the food currently grown on this planet is derived from annual crops. Accordingly, any solution or successful adaptation to climate change that hopes to feed the hungry and dispossessed over the next half century must find appropriate roles for annual crops to play in providing greater food security.

And if, for some reason, you believe that annual crops (like corn, soy, wheat, rice, and cotton) are already being genetically engineered to be "climate-friendly" and "drought-ready" enough to sustain agricultural production in the face of climate change, think again. You only need to review my friend Tom Philpott's recap of what is already happening to the three main genetically engineered annual crops—corn, soy, and cotton—and how their yields will likely fare on the most climatically challenged acreages in the United States:[2]

Back in 2008, a pair of researchers from the USDA and Columbia University shattered that comforting idea [of GMO crops increasing yields]. In a paper for the National Bureau of Economic Research, they looked at three major US crops (corn, soy, and cotton) and found that rising temperatures would indeed cause a slight increase in crop yields—up to a certain point. But when temperatures climb above that critical threshold, yields would plunge dramatically. And here's the kicker: At current levels of greenhouse gas emissions, average temperatures are expected to rise well above the critical levels identified by the researchers. As a result, they project that yields will fall by the end of this century by as much as 43 percent "under the slowest warming scenario" and 79 percent "under the most rapid warming scenario."

. . . And then in 2011, another major study, this one published in *Science* [6], found that climate change is *already* biting into yields of major crops. By 2008, global corn and wheat were down 3.8 percent and 5.5 percent, respectively, lower than they would have been without climate change, they found. For soy and rice, the good and bad effects of climate change had, to that point, largely balanced out, they found. But these crops, too, could eventually see lower yields as temperatures keep rising.

Even if we were to grant that genetic engineers may someday have more success in sustaining and enhancing annual crop yields under hotter, drier conditions than they have had in recent years, those advances will come at an extraordinarily high cost. It is estimated that it will take more than $5 million and 10 to 15 years of research and development to genetically engineer each new "climate-friendly" or "drought-ready" [sic] annual crop to be released over the next half century, and that millions more will be spent on seed patenting, grow-outs, marketing, and packaging to get each of those GMO cultivars into the fields.[3]

Nevertheless, each of those new releases may not stay in American fields for very long. Already Bt-engineered varieties of corn released less than a decade ago have had their protective mechanisms overcome by ever more pestiferous strains of the Bt-resistant western rootworm. Not only has this major pest of maize in the Midwest been spreading geographically, but higher numbers of adult rootworms feed on corn foliage in extremely dry years.[4] And the recent droughts in the United States are being linked ever more tightly to patterns associated with longer-term climate change by scientists such as Thomas Karl, director of the National Climatic Data Center.[5]

Of course, these details may potentially distract us from the seminal question regarding how we should prepare our annual croplands for drought response: *Is it ultimately more useful to invest $5 million in launching a single "drought-ready" GMO crop variety, or to invest the same amount of funds in supporting dozens of grassroots seed nonprofits, which can collectively evaluate and release thousands of already drought-tolerant heirloom vegetable and grain varieties to be cultivated on small- and mid-sized farms around the world?*

Political scientist Carol Thompson and I offered this answer several years ago in the pages of *The Economist*: "It would be far more

Regionally-adapted seeds can be obtained these days from a variety of small seed companies, nonprofits and grassroots seed libraries.

cost-effective to support local farmers in their breeding and evaluation of selected varieties already in community seed banks . . ."[6]

As Janisse Ray has described in her book *The Seed Underground*,[7] there are many novel ways that have emerged in communities over the last decade to save and exchange seeds linked to their own home ground. The majority of these grassroots and nonprofit efforts have annual budgets of less than $1 million per year, and yet many of them—like Native Seeds/ SEARCH in the Desert Southwest and Southern Seed Legacy in the more humid Southeast—maintain more than 1,000 distinct collections of drought- and heat-adapted seeds. For the cost of producing and marketing a single GMO, a dozen or more regional, tribal, or community-based seed exchanges could probably grow and offer 5,000 to 10,000 annual seed-stocks per year to farmers and gardeners in the communities they serve.

Now, more than at any other time in recent human history, we need thousands of annual crop varieties out in fields evolving and being evaluated under continually changing climatic conditions, as well as under unforeseen pressures from novel strains of diseases, pests, and weeds. But just how do we select the seeds of annual crops that are most likely to help us get through the droughts and other climatic disasters that we have begun to face with ever-greater frequency?

Seed libraries offer growers an opportunity to evaluate a seedstock and, if the variety proves well adapted to the locale, replenish it to share with others.

Permaculturist Rebecca Newburn reminds us that saving seeds is a sure way to move toward food security.

Parable

While Monsanto and other biotech companies spend millions in attempts to breed or engineer the perfect "climate-ready" annual seed crop, I was instructed 30 years ago by a Native American seed keeper not to focus on finding a silver bullet solution. This Hopi woman from Tuba

City, Arizona, redirected my attention to nurturing and keeping alive the greatest possible range of options when selecting seeds to keep back for planting in future years. She was an elder who lived not far from the Grand Canyon, near the edge of the Painted Desert, where Hopi, Navajo (Diné), Southern Paiute, and Tewa people often see one another putting up seeds for saving and sowing in the coming year. In most of these cultures, it is the woman's responsibility to both clean the seeds and select the ones that will go out for the next planting.

And so, while I watched her work through her own family's rows of dry flint and flour corns stacked with their kernels still intact on their cobs, I asked her if she selected only the biggest corn kernels of all one color for planting her blue maize. Upon hearing my question, this traditional desert elder responded to me with some simple advice:

"It is not a good habit to be too picky . . . we have been *given* this corn [by our Creator and our ancestors]—small seeds, fat seeds, misshapen seeds—all of them. It would show that we are not thankful for what we have received if we plant just certain ones and not others."

This advice didn't merely showed her gratitude toward Creation; it showed her aptitude in dealing with climatic uncertainty, which is commonplace where she lives, in a land that receives an average of only 8 inches of rain a year, and experiences wide variations in weather between consecutive years.

Recently, scientist Brenda Lin has elucidated why such a diverse mixture of seeds of an annual crop has a better chance of surviving and yielding well under changing climatic conditions than genetically uniform seeds might:[8]

- If each crop in the mix has a different shape, height, and texture, they may collectively provide **structural diversity**, which can offer pest suppression by slowing the spread of the crop's natural enemies.
- Growing "multi-line mixtures" of the same annual crop leads to disease suppression, since pathogenic fungi do not have a uniform target that they can quickly vanquish.
- Multi-line mixtures also provide a better buffer against the vagaries of climatic variability, in that some genetic lines may produce a greater proportion of the total yield in drier years, while other will perform best in moist years.[9]

The elder from Tuba City who told me not to be picky did not specifically articulate to me how her multi-lined mixture of corn seeds could potentially serve as a bet-hedging strategy in an uncertain climate; instead she took it as an expression of her gratitude that her family brought in useful harvests year after year, despite drought, heat, pestilence, and damaging storms.

On the eastern side of the Painted Desert, close to Albuquerque, New Mexico, Isleta Pueblo member Joseph Jaramillo taught me another lesson about being a traditional farmer of annual crops in this day and age: Be flexible and respond to change. Joseph has loved helping his family grow chile peppers since he was a child, when his grandfather and father grew their Isleta chiles in the full sun. But now, as the Albuquerque area has grown and warmed to summer temperatures far higher than what he experienced in his youth, Joseph plants his chiles in the cool shade of cottonwoods, which form a "fredge" on the edge of his floodplain field. Although not exactly intercropped, they still take advantage of the microclimate created by the giant cottonwoods nearby.

Principles and Premises

Following this same logic, it appears that the most effective roles for annual crops to play in enhancing the resilience of food production in a climatically uncertain future will be these:

- Multi-lined mixtures of several varieties of the same (or related) species should be planted together in the same fields, rather than in monocultural stands of a single variety.
- Some of the elements of these mixtures should include early-maturing, short-cycle crop varieties that can be planted and come to seed during brief wet seasons when soil moisture levels are temporarily adequate, thereby decreasing irrigation demand.
- These same lines and varieties of annual crops should be integrated into **intercrops** of both annual *and* perennial species, of diverse growth forms and plant families, providing **polycultures** that collectively harvest more rain and sun, and use proportionately less fossil groundwater and fossil fuel.

In short, our motto and model for using a diversity of agricultural crops resilient enough to fend off the threats of climatic disruption should be this: *No annual grown alone, no perennial left behind!*

Some varieties of amaranths and mustards qualify as quick-maturing, short-cycle ephemeral crops.

In particular, the kind of annual crops that we wish to integrate into these multi-lined mixtures and polycultures are the early-maturing, heat-tolerant varieties that ecologists call ephemeral or short-cycle drought avoiders. Why these particular extra-short-cycle crops have extra value in an era of climate change, drought, and water shortages may not immediately be obvious to everyone, but it is simple. A crop plant that matures in 45 days rather than in 60 days may require only three irrigations after transplanting out into a field, rather than four. If it uses 20 to 25 percent less irrigation to grow than its late-blooming counterpart, not only is water conserved, but the energy to pump that water is also conserved.

In particular, one kind of "annual" crop that we wish to integrate into these multi-lined mixtures and polycultures is the cultivated equivalent of desert ephemeral wildflowers. These are early-maturing, heat-tolerant varieties that avoid drought rather than tolerating it. The 60-day flour corns from the Sonoran Desert—called Harinoso de Ocho and Onaveño in Spanish—and even the Gaspé flint corn of moist temperate Quebec will begin to tassel out and produce ears in 45 days, and have dry, fully mature kernels for grinding in less than 60 days.[10] My point in offering examples from both the far Northeast and far Southwest is that every region has some early-maturing place-based crop varieties adapted to the

Drought Tolerators, Drought Evaders, and Drought Avoiders

As a food producer facing hotter, drier conditions, take a tip from desert wildflowers: There is more than one way to "skin a drought." While most seed catalogs interchange the terms **drought tolerance** and **drought evasion**, these terms are often used imprecisely to describe a whole suite of desert adaptations.[11] True drought tolerance is typically for deep-rooted, desert-hardy trees that can survive months without rains because they are tapping into underground aquifers. Just a few cultivated fruit- and nut-bearing woody perennials, like carob trees and date palms, offer true drought tolerance, for they quickly extend their roots down to tap into subsurface aquifers rather than ever suffering from drought.

However, many herbaceous annual and perennial crops do function as drought evaders in the sense that they circumvent drought. They do so by beginning their life cycle with the onset of rains intense enough to trigger germination, and complete that life cycle before the brief wet season is over. They largely avoid desiccation and drought stress by simply ripening their fruit and dispersing their seeds well before

severe soil water deficits reoccur, and so they never truly experience extended drought.[12] Many early-maturing, short-cycle vegetables and grains employ the strategies of drought evaders in desert agriculture.

Some of the best drought evaders are the desert wildflowers that ecologists call ephemerals rather than annuals, for their entire life cycles are restricted to wet periods shorter than the frost-free growing season. They do not germinate as soon as the weather gets warm; instead they sprout with the first hard rains, whether those rains arrive in April or in July. But there are also drought avoiders—perennial or annual plants that continue to live during periods of drought, but without much perceptible water loss. As their means to avoid getting stressed out during droughts, such plants prevent the decline of turgidity in their tissues by tightly closing their stomatal pores to prevent further loss of moisture from their tissues. They can also slow down their water loss by dropping leaves and sloughing off roots, allowing them to survive months without rain. Few cultivated plants other than prickly pears, agaves, and aloes are well equipped to finesse this strategy.

prevailing growing season there: Not all varieties suitable for adapting your own food production to hotter, drier conditions will necessarily come from deserts.

The second kind of annual crop that we need to seek out or select wherever we live is the kind often described in seed catalogs as "hardy in all kinds of weather" or "produces consistently under adverse conditions." As noted earlier, an uncertain climate means that we want to seek out bi-modal or quadri-modal plants[13] that are adapted to heat and cold, drought and floods, powerful winds and extended doldrums. Because global climate change has been triggering a higher frequency of extreme weather events in many prime agricultural areas around the world, we must not assume that plants adapted to hotter and drier conditions are all that we will require in the future.

The Beit Alpha cucumbers are excellent producers under the hot, dry conditions found in the Middle East and the US Southwest.

The final kind of annual food plant that will become disproportionately important over the next century is the indeterminate vine that sprawls or climbs, whether it is a pole bean, a squash vine trailing over the lip of a terrace, or a sprawling watermelon plant. That is because many of the heirloom varieties with indeterminate habits (in other words, not a compact, bushy growth form) have already been selected over decades or centuries to be suited to intercropping. If planted next to maize, millet, or sorghum, they will sooner or later twine around the stalks and climb right up them. This produces the added harvests of edible produce that agro-ecologists call the overyielding effect. **Overyielding** simply means that, for instance, the combined yield of squashes, beans, and maize in "Three Sisters" plantings grown together on the same land are often higher than what any one of these crops would produce on the same area of land if planted alone.[14]

Few seed catalogs explicitly tell us which heirloom varieties have been selected and adapted for inclusion of intercrops or polycultures. We must do our own on-farm description, selection, and evaluation of annual seed crops to determine how we can put the pieces of the puzzle back together into a functioning polyculture (see table 9-1). The crop plant guilds that function best are not simply contrived assortments of random parts. Most

Annual crop seed saving, selection, and adaptation.

of us have heard of or witnessed "companion plants" that do extremely well together. Such plant alliances can no longer be cynically dismissed as old wives' tales. We must serve as matchmakers for the plants to be intercropped in our gardens and on our farms, as my friend Suzanne Nelson did when evaluating the success of various cowpeas and tepary beans in intercrops with millet, maize, and panicgrass.[15]

Planning and Practice:

On-Farm Selection, Evaluation, and Assemblage of Quick-Maturing Grains and Vegetable Crop Guilds for Intercropping

To design functional, quick-maturing species guilds for intercropping under climatic uncertainty, there is no quick fix except to experiment with different combinations of varieties, in different spatial arrangements, with different planting dates on your own home ground. When the same crop combination is planted in two distinctively different landscapes—one with

TABLE 9-1 Early-Maturing, Heat-Tolerant, Drought-Evasive Annual Crop Varieties

CROP GROUP	VARIETY RECOMMENDED	DAYS TO HARVEST	ADAPTATIONS
AMARANTHS	Tampala	42 days	Heat-tolerant leaves for greens, stems used like asparagus
	Mayo	60 days	Heat-tolerant
	Red Stripe Leaf	28 days	Leafy heat-tolerant *gangeticus* greens from India
ARTICHOKES	Rouge d'Alger Cardoon	60 days	Adapted to Algerian climes
CHILE PEPPERS	CalWonder, Early	66 days	Loves hot days, cool nights; a sweet, thick-walled stuffing pepper
	Charleston Belle	67 days	Large red bells are heat-tolerant and resistant to root knot nematodes
	Datil	80 days	Heat-tolerant, thin-walled fruity relative of habaneros from St. Augustine, Florida
	Tabasco Short Yellow	75 days	Early-maturing, small-leaved perennial grown as annual with yellow-orange fruit
COWPEAS, CROWDERS, & LONG BEANS (FIELD PEAS)	Bisbee Black	65 days	Heat-tolerant early bloomer with black seeds
	Blue Goose	80 days	Heat-tolerant vines with large speckled purple-gray beans
	Brown Crowder	54 days	Heat-tolerant southern table pea from Mississippi; good shelled or dried
	Calico Crowder	65 days	Heat-tolerant indeterminate running vine with excellent mild-flavored peas
	California Black-Eyed Pea	55 days	Heat-tolerant, wilt- and nematode-resistant shell bean; cream with dark eye
	Chinese Red Noodle Long Bean	75 days	Heat-tolerant, indeterminate vine with small red beans; does well under adverse conditions
	Haricot Rouge Du Burkina Faso	70 days	Heat-adapted determinate bush with dark red seeds that does well in adverse conditions
	Mayo Colima Cowpea	70 days	Heat-tolerant, desert-adapted indeterminate plants
	Pink Eye Purple Hull BVR	63 days	Heat-tolerant determinate bush type with virus resistance
	Six Weeks Browneye	42 days	Quickest-maturing bush type with brown-eyed white peas
	Whippoorwill, Purple Hull	70 days	Produces consistently under adverse conditions
CUCUMBERS	Armenian Cucumber-Melon	50 days	Heat-tolerant and drought-evading vines producing light green, ribbed fruits
	Beit Alpha	56 days	Heat-adapted, mosaic-virus-resistant, pickling or salad cucumber from Israel and Lebanon
	Dekah	42 days	Early-maturing pickling cucumber from Crimea; tolerant of extreme weather conditions
	Edmondson	70 days	Drought-evading, disease- and pest-resistant heirloom from Kansas; used for crisp pickles

CROP GROUP	VARIETY RECOMMENDED	DAYS TO HARVEST	ADAPTATIONS
EGGPLANTS	Applegreen	62 days	Extra-early-maturing with apple-green fruits for northern climes
	Aswad	70 days	Heat-tolerant Iraqi variety with purple-black fruits
	Aubergine Du Burkina Faso	70 days	
	Black Beauty	72 days	Spreading, bushy plant with purplish black fruit; suited to southern climes
	Easter Egg	65 days	Early-maturing, branching small white fruits
	Florida High Bush/ Florida Market	76 days	Heat-tolerant branching plants with large purple egg-shaped fruits
	Ichiban	70 days	Early-maturing, heat-tolerant small black eggplant
	Ping Tung Long	65 days	Early-maturing, hardy, disease-resistant plants from Taiwan with long purple fruits
	Turkish Orange	65 days	Early-maturing, heat-tolerant, insect-resistant plant with red-orange round fruit
LIMA BEANS	Alabama Black-Eyed Butter	61 days	Heat-tolerant, indeterminate climbing bean
	Carolina Sieva	75 days	Early-maturing, heat- & cold-tolerant with white flat seeds
	Christmas	75 days	Tolerant of extreme heat and evasive of drought, this vine bears white & maroon seeds
	Henderson Bush	60 days	Early maturing, drought evading
	Willow Leaf	65 days	Heat-tolerant and drought evading, with narrow leaves and white or speckled seed
MAIZE/CORN	Black Aztec	70 days	Drought-evading sweet corn; white roasting ears dry to jet-black kernels
	Black Mexican	62 days	Hardy in all kinds of weather; an 8-rowed white sweet corn with kernels that dry to black
	Chapalote	90 days	Ancient heat-tolerant, slender-eared flint/popcorn
	Gaspé Flint	70 days	Dwarf, early-maturing flint for northern climes
	Mexican June	90 days	A heat-tolerant & drought-evading flour corn for late planting with delayed rains
	Reventador	80 days	Heat-tolerant and drought-evading popcorn for pinole, with translucent white kernels
	Sixty-Day Flour, Harinoso de Ocho or Maiz Blando de Sonora	55–60 days	Adapted to hot, low deserts of Sonora and Arizona; an 8-rowed flour corn
MALABAR SPINACH	Red	70 days	Heat-tolerant summer green
MELONS	Ashkabad Honeydew	80 days	Heat-tolerant indeterminate vine with sweet green flesh from Turkmenistan
	Casaba, New Mexican	90 days	Heat-tolerant indeterminate vine that bears green and orange ribbed fruit with multicolored flesh
	Persian	100 days	Heat-tolerant hardy indeterminate vine from Central Asia; fruits have fine texture and flavor

CROP GROUP	VARIETY RECOMMENDED	DAYS TO HARVEST	ADAPTATIONS
MUSTARD	Florida Broad Leaf Giant Red	23 days	Large, heat-tolerant, semi-upright leaves
	Louisiana Green Velvet Southern Giant Curled	50 days	Large, heat-tolerant, oval, bright green leaves with curled fringes
NEW ZEALAND SPINACH	Perpetual	70 days	Heat-tolerant, nonbolting summer greens
OKRA	Alice Elliot	80 days	Heat-tolerant and drought-evading Oklahoma–Missouri heirloom with green pods
	Burgundy	49 days	Early maturing, with tender burgundy pods
	Clemson Spineless	50 days	Heat-tolerant okra for the Cotton Belt, with spineless green pods
	Gold Coast	62 days	Very heat-tolerant dwarf plants with smooth green pods
	Star of David	60 days	Heat-tolerant Israeli variety with short-spined pods
ORACH	Red	37 days	Hot-weather spinach substitute
POLE BEANS	Blue Coco	55 days	High-climbing indeterminate vines; productive under hot, dry conditions, with blue snap bean pods and chocolate-brown seeds
	Rattlesnake	60 days	Heat-tolerant, drought-adapted, high-climbing indeterminate vines with streaked pods and buff seeds
	Selma Zesta/ Selma Zebra	73 days	Heat-tolerant indeterminate vines with snap beans
SORGHUM	Honey Drip Cane	110 days	Heat-tolerant, drought-evasive, early-maturing heirloom adapted to Deep South and Southwest
SQUASHES & PUMPKINS	Green-Striped Cushaw	75 days	Heat-tolerant and drought-evading striped crookneck
	Big Cheese/Segulaca	105 days	Heat-tolerant and drought-evading, fluted squat pumpkin
	Dark Star Zucchini	57–85 days	Deep-rooted, wind and drought tolerant
	Lebanese Light Green/ White Bush Marrow	40 days	Summer squash for stuffing; adapted to hot, harsh conditions
	Seminole Pumpkin	95 days	Drought-evading, humidity- and heat-tolerant heirloom from Florida; delicious flavor
SUNFLOWERS	Skyscraper	75 days	Heat-tolerant and drought evading; strong-stalked plants with large seeds
TEFF	A. L. White	115 days	Drought-hardy Ethiopian and Arabian millet for bread, beer, and injera
TEPARY BEAN	Blue Speckled	85 days	Heat-tolerant indeterminate vines with plump speckled beans
	Mitla Black	75 days	Day-length-sensitive, heat-tolerant black bean
	Pinacate	90 days	Extremely heat-tolerant mottled bean
	W. D. Hood's Brown Sonoran	60 days	Heat-tolerant and drought-evading indeterminate vine with yellow-brown flat beans
	W. D. Hood's White Sonoran	60 days	Heat-tolerant and drought-evading indeterminate vine with yellow-brown flat beans

CROP GROUP	VARIETY RECOMMENDED	DAYS TO HARVEST	ADAPTATIONS
TOMATOES	Arkansas Traveler 76	76 days	Heat-tolerant indeterminate vines with juicy rose-red fruit
	Burbank	70 days	Determinate drought-evading heirloom with high amino acid content in red fruits
	Native Sun	50 days	Determinate vines, yellow fruit
	New Zealand Paste	80 days	Heat-tolerant and drought-evading heirloom with pink flavorful fruit
	Omar's Lebanese	80 days	Heat-tolerant heirloom with sweet pink fruit
	Orange King	55 days	Determinate vine sets orange fruit under high temperatures
	Ozark Pink	65 days	Heat- and humidity-tolerant, indeterminate vines with pink, flattened, globular fruit
	Porter	65 days	Drought-evading and heat-tolerant indeterminate vine with crack-resistant red fruit
	Siletz	52 days	Disease-resistant determinate vines with red crack-resistant fruit
	Texas Wild	63 days	Low, sprawling, indeterminate vines with red-currant-like, heat-tolerant fruits
	Tiny Tim	45 days	Determinate vines with scarlet fruits
	Tropic	80 days	Heat-tolerant and disease-resistant determinate vines with red globular fruit
WATERMELON	Ali Baba	100 days	Heat-tolerant indeterminate vines also tolerate sunburn; from Iraq; fruit has delicious red flesh
	Arkansas Black		Heat-tolerant and drought-evading deep-rooted vine
	Desert King, Yellow Flesh	85 days	Heat-tolerant and drought-evading deep-rooted vine; does not sunburn
	Georgia Rattlesnake	90 days	Heat-tolerant with striped durable skin, sweet rose flesh
	Louisiana Queen	70 days	Heat-tolerant and disease-resistant striped fruit with bright red flesh
	Swift Water	68 days	Drought-evading striped-skin icebox type

sandy loam soils and cold air drainage cooling it at night, and another on clay loams staying warm under the sun all day long—you may get entirely different results. It may require that you shift planting dates for one or the other of the intercropped varieties, or change their spatial arrangements within and between rows.

In short, no two polycultural gardens or farm settings will deliver exactly the same results, even when the combinations of crops are exactly the same. That is because of the very nature of bio-complexity. Each genetically distinct variety will interact with its field's microclimate and soil in a different way, and this is described by agricultural scientists as the "genotype x environment" interaction that results in a particular *phenotypic* response in terms of flowering and fruiting time, the growth ratio of vegetative to reproductive tissues, and ultimately yield. But in an intercrop, there are also genotype-to-genotype interactions: for instance, a bean with an indeterminate habit may become vine-like and climb up one kind of slender millet plant, or it may stay bushy and close to the ground if another millet variety has enormous clusters of leaves that shade it out.

That said, there are several steps in planning that you may want to take in order to select the varieties best suited to intercropping in your area. In this exercise, we will assume that we want to find which pole bean, butter (lima) bean, or tepary bean does best when intercropped with Honey Drip sorghum, but the same process can be used to select for functionally symbiotic intercrops, companion plants, or for entire nurse plant guilds:

1. Review descriptions of different climbing (indeterminate) beans that may be adapted to your landscape or region, and consult old-timers in your neighborhood to see if there is any oral history that they have been grown nearby.

2. Obtain 100 seeds of each of three varieties of climbing beans from your neighbors, from place-based seed exchanges in your area, or from regional seed companies or nonprofits. At the same time, obtain 300 to 500 seeds of Honey Drip sorghum, and plant 200 of them in two adjacent rows (furrows) in monoculture.

3. For each bean variety, grow a quarter of the seeds (25) in full sun away from the sorghum and other grains, a quarter in alternating rows (furrows) with the sorghum, a quarter in holes in the same row planted with the sorghum at the same time, and a quarter planted in the same row five days after the sorghum has emerged. Keep the spacing of sorghum seeds about the same as it was in the monoculture planting.

4. Determine the days to first flowering, the days to first pod set, and the days to full maturity with 100 percent dried pods for each of the bean varieties in each of the four configurations: monocropped, intercropped in alternating rows, simultaneously intercropped in the same row, and staggered planting of intercrop in the same row. If some mature far later than the others, note whether they required additional irrigations.

5. Determine the sorghum seed yield in each of the three configurations that it is planted with the climbing beans, and in the monoculture as well. Determine whether there was an overyielding effect of planting

any of the beans in adjacent rows to the sorghum or in the same row by comparing the yields in the respective bean and sorghum monocultures with the yields of each of the intercrops.

6. Select the climbing bean that did best in one or more of the configurations of the intercrops, and save its seed over until the next year; save the other beans as well. The next spring, plant half of the highest-yielding bean from the previous year in the intercrop configuration where it did best, and take the other half to create a "multi-lined" mixture with the two other beans (each of them contributing a third of the total seed); plant them in the same configuration with the sorghum.

7. At the end of the second season, determine whether the multi-lined mixture outyielded the highest-yielding bean when it was planted with sorghum without the other two mixed in.

8. Continue variations of these experiments and selections each year until you have either a multi-lined mixture or a single quick-maturing, exceptionally yielding bean adapted to intercrops in your landscape.

It is clear that such a process of recurrent on-farm selection will take three to five years to have much of a meaningful long-term impact, but it is the very same process through which ancient and historic cultivators domesticated and diversified the most important food plants in the world! Although it may seem time consuming and tedious, it is rewarding. And in the face of climatic uncertainty, there is no reason to wait for someone else—especially a genetic engineer—to save us, and the food crops we wish to keep. Get started on this lifelong endeavor soon!

Experimental intercrops of lettuce, kales, mustards, and peas can help determine optimal spacings for polycultures.

Chapter Ten

Getting in Sync

*Keeping Pollinators in Pace and in Place
with Arid-Adapted Crop Plants*

Warm-Up

What would you do if you showed up for a date, but after a few hours of waiting, you realized that your partner had arrived several hours earlier and then left? Well, that's one of the potential scenarios for how climate uncertainty is likely to affect crop plant–pollinator interactions in the near future, if it has not already begun. The polite term for this temporal mismatch is *asynchrony*, but the colloquial expression is *getting stood up*!

There is growing evidence that many pollinators and plants are being triggered into earlier but not necessarily synchronous activity by the same temperature shifts associated with global warming. However, each partner may be responding differently to these shifts in space and in time, so much so that they are increasingly vulnerable to "ecological mismatches."[1] Even if plants and pollinators do respond to the same temperature cue, the strength of the response might differ.[2] Certain pollinators and their host plants may have successfully interacted with one another for centuries, but now each of their **phenologies** or seasonal activities has begun to shift at rates somewhat independent of their partner's. From a data bank of 1,420 kinds of pollinators that are known to visit some 429 kinds of plants, it has been predicted that climate-driven changes in flowering times will reduce floral resources for at least 17 percent and perhaps as many as half of all pollinators, resulting in diminished nutritional diversity within their diets.[3]

While it has already been amply demonstrated that a few pollinators such as bumble bees are getting out of step with certain wildflowers, there is less consensus on whether the flowering of many crop plants has become increasingly asynchronous with their primary pollinators.

Pollinators are attracted to mixed stands of perennial aromatic herbs at the Generalife gardens in Granada, Spain.

Nevertheless, it is worrisome that wildflowers such as glacier lilies are now flowering two to three weeks earlier than they did a couple of decades ago, for this may foreshadow what might soon happen to at least some food crops. At certain sites in the Rockies, bumble bees are waking up and emerging from mountain meadows later than they are needed to perform the bulk of pollination required by glacier lilies. Across the continent in Maryland, nectar flows from the flowers of trees are beginning a month earlier than they did in the past, as a result of warmer winter temperatures. At my own office on Tumamoc Hill in Tucson, hibiscus flowers are, on average, opening up and flowing with nectar a month earlier than when the first desert ecologists recorded their flowering in the early 1900s. But averages don't tell the entire story; compared with the May 23 flowering at the onset of data taking by my predecessors around 1906, hibiscus flowers are sometimes opening up *70 days earlier* than they did a century ago, perhaps due to the urban heat island effect exacerbating the influences of global warming.

At the same time, other reports suggest that the suites of available pollinators—and *not* the flowers themselves—show the most dramatic shifts in timing or in northward movements.[4] Some of them are moving up mountainsides as much as 1,800 feet higher than in the past, while others are emerging several weeks earlier in the year, even as their associated flowers are blooming for shorter and shorter periods each season.[5] In still another field report from the northeasternmost reaches of North America, scientists found that, over the past 130 years, the average springtime emergence of 10 different bee species has advanced as much as 9 to 12 days. Most worrisome is that much of this advance has taken place since 1970, and it closely parallels global increases in temperatures.

If we take some hints from what is already happening to the pollinators of plants native to wilderness landscapes, we can imagine that several different scenarios may eventually occur in agricultural landscapes:

- Due to warmer winters and longer growing seasons, a crop plant will break bud and flower in advance of most bees emerging at the same site, leading to either reduced fruit and seed set or starved bees.
- Due to warming on site, nearby, or farther south along a migratory corridor, pollinators arrive and then leave in advance of flowering, resulting in no seed or fruit set at all.
- Due to different triggers activating both pollinators and crop plants in the spring, there is only partial overlap in the timing of flowering and pollinator availability, leading to reduced fruit and seed set, as well as limited nutritional rewards for the pollinators.
- Due to extremely hot springs and summers, bees or other animals pollinate the crop's flowers, but the flowers still abort due to heat stress.
- Due to unfavorable weather that limits their nutrition and reproduction year after year, more and more pollinators are starved to death, their populations decline, and their species become endangered.

If you are skeptical that these scenarios may currently or eventually apply to crop pollinators, look again at recent reports in science journals and newspapers.[6] Bumble bee species known to pollinate both wildflowers and crop plants are declining in several regions of North America,[7] with some of them becoming so rare that they have been nominated for the US Endangered Species List. The abundance of particular bumble bees has declined by as much as 96 percent in just two decades, and their geographic ranges have shrunk by 87 percent. While climate change is not the only factor causing their declines, it is one more contributor to "the perfect storm" that has ravaged pollinator abundance over a single human generation.

Up until the last two decades, the overall decline in agricultural pollinators was largely independent of climate change, but the various pressures on bees have certainly hurt the production of certain crops already, in terms of diminished yield and/or rising costs of production.[8] While the number of managed colonies of honeybees in the United States peaked at 5 million in the 1940s, we've lost at least 59 percent of that number since then, and climate change is only one of the many insults they have suffered. Introduced parasitic mites, fungi, and viruses, not just neonicitinoid insecticides, have taken their toll on honeybees.[9] In the 2009–10 winter season, scientists confirmed for the first time that honeybee numbers are globally dwindling, with 24 countries suffering higher annual losses than they had ever reported before. Since the appearance of **honeybee colony collapse disorder** in North America in 2006–07, beekeepers have been suffering winter losses of hives that have ranged from 20 to 36 percent per year. With losses in other countries just as severe as those in the United States, it is unlikely that American beekeeepers will be able to continue to import managed colonies from other continents as cheaply and as casually as they have in the past.

Bee scientists warn that losing a roughly a third of the carryover population of bees from year to year can only go on for so long before honeybee scarcity will put the price of renting managed colonies out of the reach of many farmers. Already, California almond growers have seen rental prices jump from $35 a colony in the 1980s to over $150 a colony today. That's because the almond crop is 100 percent dependent on bees for pollination, and so enough bees are needed to fully pollinate California's 750,000 acres of almond orchards in mid-February, or else growers will suffer dramatic yield losses. If a beekeeper brings in 20,000 colonies to pollinate a 1,000-acre almond orchard, he or she can gross $300,000 in a matter of three to four weeks (only 5 to 10 percent of which becomes net income). At the same time, almond growers are so dependent on these mobile beekeepers to ensure a lucrative yield that one-fifth of their entire operating budget is for pollination services![10]

It came as no surprise to seasoned beekeepers that their honeybee colonies would suffer as a result of the 2011–12 drought that hit two-thirds of North America.[11] With less nectar and pollen to forage, it is clear that less honey was produced, and likely that pollination services were diminished. It prompted the *New York Times*'s Harvey Morris to repeat an adage attributed to Albert Einstein: "If the bee disappears from the face of the earth, man would have no more than four years left to live."

Parable

As food producers, how do we stay alert to the possibility of mismatches between the flowering of our food crops and the pollinators that serve them, when these interactions are so complex that scientists have difficulty predicting their outcomes? At one point in my life, I despaired that I had so little capacity to recognize and respond to climatic shifts happening even in my own field near Flagstaff, Arizona. The food crops

I had planted—independent of what was happening to their pollinators—seemed so vulnerable to the vagaries of drought and temperature extremes that I could not imagine how anyone could manage the intricate interactions among them.

But then, by chance, I learned a simple lesson from a Seri Indian naturalist who had lived on the desert coast of Mexico's Sea of Cortés his entire life. Humberto Romero Morales had worked as a fisherman and sea turtle hunter early in his career, but had left those occupations behind to identify and monitor the plants used by desert tortoises and bighorn sheep for his tribe's fledgling wildlife management program. He was still a keen observer of sea turtle behavior, but now more as a conservationist than as a traditional hunter. At the moment, he was guiding me and a number of environmental science students in a spring survey of desert tortoise abundance in the coastal range just above his home.

As all of us were walking along, 5 yards apart from one another, looking for desert tortoises along a half-mile-long transect, Humberto stopped us for a moment to point out something he had just observed.

"Look," he exclaimed to me and the other surveyors, "the first ocotillo flowers of the season have begun to bloom on the rocky ridge above us. That means that the first young migratory sea turtles should be arriving near shore below us, coming up the coast from their winter stayovers to the south!"

One of the wildlife biology students who accompanied us rolled his eyes and blurted out, "That makes no sense at all. How does a green sea turtle know or even care when a desert ocotillo plant begins to flower? One of them is on land, the other in the sea . . . their lives are completely unconnected!"

When Humberto realized what the student was implying, he turned toward us and gave us a lesson in ecology that none of us had ever previously considered:[12]

"What you all don't understand is that the sea turtles aren't looking at ocotillos to guide them into migration, but they are both triggered by the same forces that underlie everything that happens here . . . We are using one as a visible indicator of what may be happening to the other, because it is easier to see. But both are responding to the same fundamental shifts in weather."

It seems that pollination ecologists have begun to realize the very same principle: While plants and their pollinators to some extent have different responses to climate change, "the phenology of plants and trap-nesting bees and wasps is regulated in similar ways by temperature."[13] Humberto was not arguing cause-and-effect relations between the timing of ocotillo flowering and that of sea turtle migrations; instead he was suggesting that the timing of the two events was correlated because they are being triggered or regulated by the same underlying causes.

It is not currently possible to precisely predict for your particular foodscape how much each food crop and its various pollinators will shift their timing and whether they will get substantively out of sync. But if you begin to pay attention to the trends in temperatures while making observations in both warm and cool years, you may begin to see how the underlying causes of plant and pollinator activity shifts will play out where you are producing food.

Principles and Premises

Most fruit and vegetable crops require pollination by animal vectors at levels sufficient enough to set seed, ripen fruit, and allow them to mature as fully formed, nutritionally rich foods. An apple tree that lacks adequate pollination may lack its full array of 10 germinable seeds per fruit, and its fruits may be shriveled or misshapen. In fact, at least some populations of most plant species in the world currently suffer from inadequate pollination due to pollinator scarcity.

Bob Duncan keeps nest boxes for the blue orchard bees (*Osmia lignaria*) that pollinate many of his 200 varieties of fruits near Victoria, British Columbia.

In attempting to avoid such problems—by adapting to climate change through maintaining enough pollinators in your foodscape—there are four general rules of thumb to keep in mind:

1. It is best to invest in pollinator-habitat-enhancing practices that offer redundancy through the presence of a diversity of wild and domesticated pollinators, rather than simply investing in efforts to keep around a single pollinator species—whether it is a honeybee, bumble bee, butterfly, or bat.

2. That said, enhancing a few pollinators that are uniquely matched to your crop mix—for instance, blue orchard bees for an orchard of mixed fruits, alfalfa leaf cutter bees for a pasture of leguminous forages, monarchs for milkweeds and their kin, or squash and gourd bees for your pumpkin patch—will never hurt. Honeybees, of course, are terrific generalists and remain valuable for diversified CSA farms and many other settings.

3. The ideal situation to achieve is one in which wild or cultivated nectar plants are planted early enough so that they may begin to bloom on your land in advance of your crops, and maintain enough floral resources to keep a critical mass of bees in your field or orchard through the collective flowering times of all your crops.

4. Pesticides and herbicides are likely to disrupt wild pollinators with body sizes smaller than honeybees even more than they disrupt managed honeybee colonies.

It has become increasingly evident that the 4,000 to 6,000 species of native bees and other crop pollinators in North America provide a "bet-hedging strategy" or natural crop insurance against both the loss of honeybees and the risks of pollinator disruptions due to catastrophic climatic events.[14] When a team from Rutgers University investigated whether the potential loss of honeybees could be compensated for by native, wild bee species, they found that populations of native bees alone were sufficient to sustain adequate pollination on 90 percent of the 23 watermelon farms they surveyed in Pennsylvania and New Jersey. The field researchers felt confident that, at least in the Mid-Atlantic states, the pollinator redundancy offered by native bees will be sufficient to buffer farmers from potential declines in agricultural production that may result from the ongoing national and global losses among honeybees.

But just how do we keep native wild bees in place and in pace with the flowering of crop plants? I conducted field studies as part of the Forgotten

Bob Duncan shows off his nest boxes for blue orchard bees.

Pollinators Campaign's teams of researchers in four parts of North America, and we determined that the close proximity of healthy wild habitat adjacent to fields and orchards was far more effective than sowing rows of pollinator-attracting forage plants in the fields, or constructing bee boxes or other nesting sites in an orchard.[15] We found that in riparian habitats adjacent to a mix of forage pastures, vegetable fields, and orchards, both migratory pollinators and other invertebrate pollinators in fragmented habitats benefited from this type of linkage. Our initial results have been validated by more recent studies, which also recommend that the width of these corridors of agro-habitats should be greater than one home range. They also provide an economic incentive to preserve wildland habitats in larger food-producing working landscapes.[16] In Canada, for example, it has been demonstrated that higher yields associated with higher populations in "bee meadows" adjacent to fields makes it cost-effective to convert a third of the acreage on a farm in field crops back to wildland habitats or diverse cultivated meadows.

Squash and Gourd Bees: Allegiant and Economic Pollinators

Over half of my life, I've delighted in finding native squash and gourd bees sleeping in squash blossoms at night, then beginning their pollination work shift well before dawn. The males and females find each other in the squash blossoms, and mate as they pollinate! If you might think that such frolicking reduces their effectiveness as pollinators, guess again. Early in the season, males visit virtually every blossom that opens in a pumpkin patch or squash field. Along with bumble bees, they are far more effective than honeybees in pollinating these crops, since they seldom if ever visit flowers other than cultivated cucurbits (squashes, pumpkins, or gourds) or their wild relatives such as the buffalo gourd, coyote gourd, or Okeechobee gourd.

Where you can find the regular presence of at least one species of squash and gourd bees such as *Peponapis pruinosa*, there may be no need at all to rent honeybee colonies for your pumpkin patch. My friend Jim Cane has estimated that in one particular field of 90,000 squash plants, close to a million squash and gourd bees effectively accomplished all the pollination services that needed to be done. That was perhaps bittersweet news to the farmer who cultivated that site, since he had already paid $25,000 for the rental of honeybee colonies to ensure a good crop![17] The wild native squash bees likely completed most, if not all of the pollination before the honeybees even awakened in the morning, since the squash bees begin their activity as early as 4 AM, visiting flowers just as they open for the day.

Pollinator habitat can be developed along barrier fences on terraces.

Pollinators benefit from field-edge and between-row plantings of wildflowers in this orchard in Sonora, Mexico.

However, immediately along the edges of fields and orchards—in and near the hedges and fencerows we call *fredges*—there is much that food producers can do to accommodate a wide variety of pollinators, especially bees. Different kinds of bees nest in barren but often unplowed ground; in hollow stems or pithy sticks; or in dead trunks, woody flower stalks, or fence posts (see table 10-1).

In addition to taking care of roosting, nesting, and (for butterflies) larval host plants, the key issue is to ensure pollinator diversity and abundance. The availability of sequentially flowering nectar plants in these bee meadows and other wildland habitats can often ensure this abundance and diversity, which in turn favors better rates of pollination and fruit-set.

One of the best means of making a place for on-farm pollinator diversity is to design fredges and **vedges** (plantings of pollinator-attracting vines that trail over a fence) to provide food and shelter for both localized center-foraging pollinators and long-distance migratory pollinators.[18] These same hedges or fencerows tend to harbor beneficial insects other than pollinators as well.

A hand-constructed nest box for native bees at Las Milpitas in Arizona.

TABLE 10-1 Habitat Requirements for Different Nesting Guilds of Crop Pollinators

POLLINATOR GUILDS	NESTING & ROOSTING HABITATS	TAXONOMIC GROUPS
Leaf cutter bees	Open habitats where they can dig nest holes on barren ground	Megachilidae, esp. *Megachile*, *Heriades*, and *Ashmeadiella*
Mason bees	Preexisting cavities, pithy or hollow plant stems, small rock cavities, abandoned insect burrows, or even snail shells	Megachilidae, esp. *Osmia*
Carpenter bees	Woody substrates like broken or hollow twigs, or open habitats where they can dig nest holes	Two genera within Apidae (*Xylocopa* and *Ceratina*) and one within Megachilidae (*Lithurgus*)
Social bees (including honeybees & bumble bees)	Preexisting cavities	Apidae: honeybees, bumble bees, and stingless bees
Squash & gourd bees	In deep, moist soil on floodplains or other flats where barren, sparsely grassy, or branch-covered ground occurs	Two genera within Apidae (*Peponapis* and *Xenoglossa*)
Butterflies	Specific larval host plants are required for each kind of butterfly, while sites for nests and roosts are more generalized	Many nectar-feeding genera and species within Lepidoptera
Hummingbirds	Low to medium branches of shrubs and trees in habitats with ample "woolly" or pilose plant tissues for nest-building	Trochilidae

Pollinator gardens can be planted inside stone walls or living fences (fredges and vedges) to attract bees to crops.

Planning and Practice

In preparing to construct and plant a vedge (vining fence/hedgerow) for pollinators, there are several important steps before selecting and propagating plants to climb up onto and trellis along a fence. As a recent UN Food and Agriculture report on climate change and pollinators suggests:

> The first step is to identify . . . environmental cues controlling the phenology of important pollinators . . . [This] might include maximum daily temperature, lack of frost, number of degree days (number of days with a mean temperature above a certain threshold), day length and snow cover. It is also important to record climatic data in the area where the crops [themselves] will be planted (e.g., average temperature, precipitation, snow cover) to identify other areas where the results [have historically been] similar.[19]

Pomegranate flowers offer nectar to New World hummingbirds and Old World sunbirds.

The reason that we should look to see how our current conditions are similar to those historically found in other areas is that this comparison may help us choose pollinator-attracting plants from those areas that may not currently be abundant in our own. As climate change accelerates, it is valuable to select pollinator-attracting plants that will meet the warmer (or at least more extreme) conditions that may become more prevalent, rather than those appropriate for the historic conditions in our foodscape.

Because different sites in various regions will clearly demand different sets of pollinator-attracting plants, it is worth your while to review the "regional plant lists for pollinator gardens," available online from the Xerces Society, or printed in its published guide, *Attracting Native Pollinators*.[20] But because conditions may shift radically over the next two decades, look to native plant mixes that currently grow at lower elevations or miles to the south of your current location. If the vedge or fredge is already in agricultural lands where you want perennial cultivated plants that attract a great variety of pollinators and hold the soil in place, though, select some of your species from table 10-2. It recommends some wide-ranging domesticated crops for use, but it will require you to find a variety of the species of interest that is best suited to your climate.

This sculptural fence at Sooke Harbor, near Victoria, Canada, is edged by pollinator-attracting herbs and edible flowers.

TABLE 10-2 Cultivated Perennial Species Recommended for Pollinator-Attracting Fredges and Vedges

COMMON NAME	SCIENTIFIC NAME	HABIT & ORIGINS
Alfalfa/lucerne	*Medicago sativus*	Perennial forage from Southwest Asia
Beebalm/bergamot/Oswego tea	*Monarda didyma*	Aromatic perennial from Mediterranean
Buckwheat	*Fagopyrum esculentum*	Annual from China
Fennel	*Foeniculum vulgare*	Perennial from Mediterranean
Hyssop	*Hyssopus officinalis*	Erect perennial herb from Southeast Europe/Mediterranean
Marjoram	*Origanum majorana*	Aromatic perennial from Mediterranean/Levant
Peppermint	*Mentha x piperita*	Aromatic perennial herb, now ubiquitous
Rosemary	*Rosmarinus officinalis*	Aromatic evergreen subshrub from Mediterranean
Sage	*Salvia officinalis*	Aromatic perennial subshrub from Mediterranean
Spearmint	*Mentha x spicata*	Aromatic perennial herb from Europe and Levant
Sunchoke	*Helianthus tuberosus*	Tall perennial with rhizomes from North America
Thyme	*Thymus vulgaris*	Aromatic perennial herb from Mediterranean
White clover	*Trifolium repens*	Perennial clover from Mediterranean

Let's walk through the stages of planting a vedge of perennial vines that climb and sprawl over either a traditional wooden **retaque fence**, or a living ocotillo branch fence, both common in Mexico. A retaque fence consists of mesquite or acacia branches 4 to 5 inches in diameter, horizontally stacked between sets of mesquite fence posts that are dug into the ground about a foot apart; the paired fence posts form a chute within which the other branches can be stacked. Ocotillo fences are made by cutting living but dormant wand-like branches off *Fouquieria splendens*, a wand-like desert plant that grows from Texas to California and south into Mexico. Once rooted in the soil, its pruned branches can be top-watered or misted until they sprout leaves, which then stimulates rooting. Ocotillo fences have four to six branches per foot and are wired at the top and bottom between fence posts that are dug in every 5 to 8 feet. They will sprout leaves after each rain, and produce clusters of red flowers, which attract hummingbirds and bumble bees.

To construct a vedge of pollinator-attracting vines, follow these steps:

1. Select a site on the edge of a field or orchard where you do not wish deer, pronghorn antelope, wild pigs, peccaries, or other large herbivores to enter. Ensure that the soil does not have layers of bedrock or caliche near the surface, and, if possible, place it where you have a relatively porous sandy loam.

Stacked mesquite fencing immediately behind cattle is ideal for creating a vedge, or vining hedge, for pollinators.

2. Construct the fence of your choice (retaque, ocotillo, or otherwise) so that there is at least a 3-foot strip of arable land on the edge of it suitable for planting perennial vines. Dig a 2-foot-deep trench, 18 inches out from the fence, and fill it with compost, mulch, effective microbe solution, and/or mycorrhizal inocula.

3. Every 3 feet, plant a perennial vine (see table 10-3 for selections), ensuring that you have at least three individuals per species along the fence. If the transplants are already vining out, train and trellis them up into the fence.

4. If available, lash last year's flower stalks of century plants (*Agave* spp.) and desert spoons or sotol (*Dasylirion* spp.) to the fence posts, to serve as nesting sites for carpenter bees.

5. Top-water, drip-irrigate, or mist the vedge once every week until established, and occasionally sickle away or prune any plants competing with the pollinator-attracting vines.

TABLE 10-3 Desert-Adapted Herbaceous Vines for Pollinator-Attracting Vedges

SCIENTIFIC NAME	COMMON NAME	POLLINATORS
Apodanthera undulata	Coyote melon	Butterflies
Aristolochia watsoni	Southwestern pipevine	Pollinated by midges but larval host plant for pipevine swallowtails
Clematis drummondi	Virgin's bower	Bumble bees, butterfly larval source
Clitoria mariana	Butterfly pea	Butterfly larval host plant
Cucurbita digitata	Coyote gourd	Squash and gourd bees, bumble bees, honeybees
Cucurbita foetidissima	Buffalo gourd	Squash and gourd bees, bumble bees, honeybees
Galactia wrightii	Wright's milkpea	Butterfly larval host plant
Lachnostoma arizonica	Arizona rib-pod milkweed vine	Butterfly larval host plant
Macroptilum atropurpureum	Siratro wild bean	Mostly honeybees, but a wide variety of Andrenid, Apid, Halictid, and Megachilid bees
Marsdenia edulis	Milk vine	Flies
Morandia antirrhiniflora	Snapdragon vine	Hummingbirds
Parthenocissus inserta	Arizona creeper	Honeybees and butterflies, and moth larval host plant
Passiflora arizonica, byronoides, & mexicana	Native passion vines	Pollinated by carpenter bees and honeybees but also excellent butterfly larval host plants
Phaseolus coccineus	Runner beans	Bumble bees and honeybees
Rhynchosia senna	Rosary bean/sanipusi	Carpenter bees
Sarcostemma cynanchoides	Climbing milkweed	Monarch and striated queen butterflies

Huachuca agaves in a clonal population provide enormous quantities of nectar and pollen to migratory pollinators such as bats and hummingbirds.

A Vedge for All Seasons on the Edge of the Desert

For our location at 4,000 feet on the border between Arizona and Sonora, Mexico, we have selected species that currently occur at the upper elevational edge of the Sonoran Desert at 3,000 to 3,500 feet in elevation. Because we required a barrier to keep javelinas or wild peccaries from foraging on our terraced prickly pear cacti, we constructed a retaque fence of stacked mesquite and pecan branches on the bottom edge of a ridge above our entrance road. Then, as part of a Borderland Habitat Restoration Initiative workshop funded by a USDA Western SARE grant, we transplanted 22 herbaceous perennial shrubs and vines on the interior of the fence. Our goal was to let them climb up the 4-foot fence, then sprawl down over it to help hold the eroding bank cut above the road in place, since many of these plants will root wherever the nodes of their vines trail across moist soil. Another goal was to provide a regular stream of pollinators, not only for my orchard of 70 varieties of edible succulents, fruit, and nut trees, but for 60 acres of mixed vegetable crops on the Native Seeds/SEARCH farm, located just 100 yards downhill from our orchard.

This last year, we had plants flowering in the "Vedge on the Desert's Edge" from March 20, when we transplanted most of the vines and subshrubs along the retaque fence, all the way through October 10, when the first light frosts hit our valley. Along the way, we became the first Certified Bee-Friendly Farm in southern Arizona through Partners for Sustainable Pollination and are now beginning to reap the fruits of that effort. In fact, some of our fruits won a Santa Cruz County Fair blue ribbon—an award that the bees, wasps, and butterflies truly deserve! Perhaps more importantly, as a result of our workshops, at least ten other farms and gardens in Arizona have joined the Bee-Friendly ranks.

Creating Your Own Sowing Circle

A Resilient Food System Takes More than a Farm

Sowing Circle

We sow these words
Like seeds to the wind
Hoping they will find
The earth they need
To endure the drought
And bear the fruit
For us to eat in a future
More unfathomable than ever.

After dawn, we moved stone
Next to stone next to stone
To terrace a slope so
To keep the earth in its place so
That it could hold water
Even when everything around us
Feels dry as a bone.

Such measly gestures
Done with rough hands
Burning minds, pumped-up hearts
Cannot break any drought, forestall any flood
But it may keep us rooted
And maybe even fed
In the face of such uncertainty.

—GPN, October 2010, Patagonia, Arizona

The stories in this book have been sown like so many seeds onto an American earth that has seen its soil deteriorated, its waters desiccated, and its natural abundance depleted and nearly destroyed by decades of misguided use. We need only to recall the words of Masanobu Fukuoka, the great Japanese "do-nothing" farmer, for they remind us how much of the world sees what we have done to this continent, and how future generations may see what we have done in less than a century and a half of growing food with fossil fuels and fossil groundwater: "I was expecting the American continent to be a vast, fertile green plain with lush forests, but to my amazement, it was a brown, desolate semi-desert."[1]

But this is also a book about how the time-tried practices of traditional desert farmers from many parts of the world can help us deal with and adapt to climatic uncertainty. Their knowledge and wisdom can clearly offer hope to those who are trying to produce food in true deserts, but they can help farmers and gardeners in temperate and tropical regions as well, to avoid desertifying their food-producing landscapes. I would further argue that if we can adapt, refine, and fine-tune some of these desert farmers' practices to the specificities of our own foodsheds, our own agrarian landscapes will look less like stereotypical deserts and more like productive oases, where the constraints of hot, dry conditions are respected and worked with rather than ignored. In short, we will be able to "meet the expectations of the land" as Thoreau admonished us to do. That means that we must fit our crops and practices to the current and future realities of the environment in which we live, rather than trying to remake the environment (at enormous cost) to fit the crops we wish we could grow.

Most of this book has focused on what we can do as food producers to reduce our ecological foodprint, and grow nutritious fruits, vegetables, and cereals without further contributing greenhouse gas emissions to global warming. But it is critical for all of us who eat (that may include you!) to remember that much of the carbon and water foodprint of Americans, Canadians, and Mexicans comes from what happens to our harvest once it leaves the farm-gate for the market, cafeteria, or processing plant. Estimates vary, but the fossil fuel (and other energy) used in growing out food on farms and ranches accounts for roughly one-fifth of all the energy use embedded in our food value chain by the time a fruit, vegetable, meat, or beverage reaches your plate.[2]

Let me repeat that: *Farming accounts for only a fifth of the energy used to get food to our plates.* The following pie chart shows all of the energy expended in order to bring 1 calorie of food to our table, according to researchers Heinberg and Bomford, based on University of Michigan analyses.

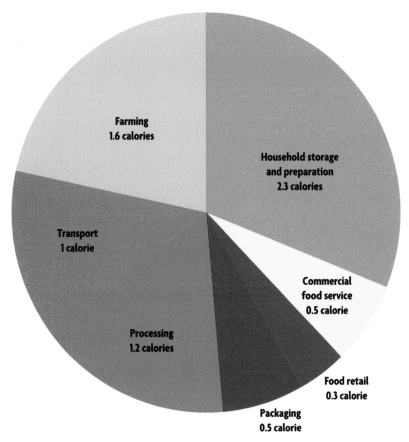

Farming
1.6 calories

Household storage
and preparation
2.3 calories

Transport
1 calorie

Commercial
food service
0.5 calorie

Processing
1.2 calories

Food retail
0.3 calorie

Packaging
0.5 calorie

Energy expended in producing and delivering one food calorie. Approximately 7.3 calories are used by the US food system to deliver each calorie of food energy. Farming accounts for less than 20% of this expenditure, but still consumes more energy than it delivers. SOURCE: RICHARD HEINBERG AND MICHAEL BOMFORD, 2009. THE FOOD AND FARMING TRANSITION: TOWARD A POST-CARBON FOOD SYSTEM. POST CARBON INSTITUTE, SEBASTOPOL, CALIFORNIA.

What this pattern suggests is that much of the ecologically sensitive work done by food producers trying to make the most of sun and rain in their farms, ranches, and market gardens is *undone* by what happens on the highway and in the cannery, the feedlot, the cafeteria, the fast-food restaurant. Farmers and ranchers in the United States today gain only 6 to 18 cents on every food dollar spent by consumers for good reason: 8 to 20 other hands may touch that food once it leaves the farm-gate, trimming and cleaning it, bulking it for distribution, shipping it, repackaging it, misting it every 20 minutes, placing it in heat-sealed packages or in displays, or homogenizing it with dozens of other ingredients to provide microwave-oriented Americans with the thawed remains of their frozen processed foods. Although such foods are typically called *value-added products*,

TABLE A-1 How Land Health, Human Health, and Community Economic Health Can Be Restored to Create Resilience in the Face of Climate Change

LAND HEALTH	HUMAN HEALTH	ECONOMIC HEALTH
Restoration of **nature's services** (water and pollinator flows) upon which resilient food production depends	Regeneration of immune defense system (probiotics and antioxidants) upon which individual health depends	Resurgence in the use of locally available capital (for equity and loans) upon which community economic health depends
Landscape-level rebuilding of regional infrastructure for sustainable food production, transport, and processing	Community-level rebuilding of diverse disease prevention and health care options accessible to all classes	Regionwide coordination of investors willing to help jump-start and sustain farming, food, and health micro-enterprises
Minimizing dependence upon fossil fuels, fossil groundwater, and petrochemical-based fertilizers	Minimizing dependence upon silver bullet surgeries, antibiotic pharmaceuticals, chemo- and radiation therapies	Minimizing dependence upon multinational banking institutions and governments for loans and equity shareholders
Nurturing a high diversity of soil microbes and beneficial insects in fields and orchards and compost pits	Nurturing a high diversity of microbes in human guts, kitchens and breweries, and fermentation vats	Nurturing a high diversity of local participants in micro-enterprises and models for innovation/ demonstration
Rescuing and renewing traditional knowledge and seeds through open-source access to those adhering to ethical protocols	Rescuing and renewing traditional knowledge about home remedies, home fermentation, and suppression of allergies with local organisms	Rescuing and renewing traditional knowledge about local trading of work and materials, bartering and sharing tool and seed libraries

some of them should truthfully be called *value-diminished products*, for much of the sun, rain, and nutritive value have been squeezed out of them, while fossil-fuel-mediated additives and packaging have increased their total greenhouse emissions fivefold.

Because of such realities, all of us must begin to "think like a food-shed" rather than thinking that however much sustainability we weave into our farm, garden, or ranch, what we do on our own land is enough. As Anna Lappé reminds us, there are many other ways we can impact "the climate crisis at the end of our forks."[3]

From where I sit, it seems that we need to "ferment" a new revolution, not only about how we care for the land, but how we care for our health and for our local economies (see table A-1). The steps suggested on these three pathways to restoration will not only enhance the food-producing capacity of our land, but create less waste in our food system and redirect the community economies in which our food production is embedded. But we must pay attention to each of the three legs of this stool for seating restoration, regeneration, and resilience—for without even one of these three in place, the stool will topple.

To do such necessary work, we must join together with our neighbors—adjacent landowners, river keepers, food chain restorationists, educators, and consumers—to create local "sowing circles" that can redesign and stitch back together the foodsheds we rely upon. We must lower the fences between our "territories" and embrace the tenets of collaborative conservation that thousands of Americans have pledged to support through signing "the radical center manifesto."[4] In a world where collaborative conservation of natural resources and cultural values is manifested in our foodsheds, we may converge upon the vision that agrarian poet and naturalist Rainier Maria Rilke expressed nearly a century ago:[5]

All will come again into its strength:
The fields undivided, the water undammed,
The trees towering and the walls built low.
And in the valleys, people as strong
And as varied as the land.

Drawing by Paul Mirocha.

Glossary

Agro-biodiversity is the accumulation of food-producing varieties and species in human-managed habitats as the result of 10,000 years of natural and cultural selection and informal conservation.

Assisted migration is the intentional conservation practice of transplanting individuals or populations of a rare species from their threatened habitats to another site or sites with conditions more optimal for their growth, reproduction, and survival. The intent is to establish a refuge where the artificial population will be safer from the impacts of climate change or other human-induced pressures than it might be in its current home range. Although it is still a controversial practice in the conservation of wild species—even as it was when all California condors were removed from fragmented wild habitats in Southern California to be captively bred then reintroduced elsewhere—it may be a more viable strategy for safeguarding place-based food crop varieties, particularly if the donor communities and recipient communities of farmers maintain ethical agreements and food exchanges.

Balcony or **window-box terraces** are raised level places for planting along hillside, mountainside, or ridgeside drainages. Terraces are usually fortified by stone walls, with organic matter and exogenous soil media used to fill in the space behind the wall to create a level surface area and the capacity to collect storm runoff to store in the soil.

Bench terraces, especially the Zingg types, are designed for achieving maximum moisture conservation for plant production behind a terrace wall or embankment. An earthen embankment like that of a ridge terrace is typically constructed, rather than a stone or cement wall. After its construction, the portion of the terrace immediately above the ridge is leveled to form a flat bench so that stormwater can be retained and infiltrate into the soil.

Bimodal or **quadri-modal** plants are those adapted to tolerating two or more abiotic stresses (fire, flood, wind, droughts) that are threatening to become more common with accelerated climate change. Although British horticultural writer Alice Bowes suggests that a greater focus on bimodal plants is needed if we are to adapt to more frequent floods and droughts linked to climate change, we actually need to select species or ecotypes that can tolerate as much as four kinds of weather extremes—hence the modified term *quadri-modal*. Their survivability is to a large extent dependent upon versatile plant architecture, including the pliability of branches, deciduousness of leaves, and capacity of roots to slough off root hairs or survive waterlogging.

Biochar is the name given to a special category of charcoal whose particular physical and chemical properties make it safe for storing carbon in agricultural soils. As a soil amendment that has been used in the Amazon Basin for millennia, biochar came under investigation as we attempted to understand how indigenous South Americans sustained soil fertility and food production at the same sites over time. Formed by the slow pyrolysis of organic biomass in oxygen-depleted environments, biochar has been found to improve both the water-holding capacity and nutritional composition of plants in certain tropical or temperate soils where charred woody plants, bones, and pottery shards have been intentionally deposited by decades of farmers. Its value in other soils, such as those typical of deserts, remains unproven; other traditional techniques such as the accumulation of flood-washed organic detritus may perform the same functions for desert crops.

Biomimicry is the process of using nature as a model to design novel products, processes, or systems in order to solve problems currently affecting human societies and their survival. Although the term was coined decades ago from *bios*, life, and *mimesis*, to emulate, mirror, or mimic, it was more recently popularized and articulated into a formal design process by Janine Benyus in a book by the same name. If nature has generated design solutions for life-forms over billions of years, than the study of biomimesis is the pursuit of applying such design solutions inspired by nature.

Bokashi fermentation is an alternative to conventional composting by activating an anaerobic fermentation process on organic materials to transform them into useful soil for food production. The term *bokashi* is a Japanese word meaning "fermented organic matter." With the bokashi method, inocula of effective microbes can be sprinkled over food waste placed in low-oxygen containers or enclosures. These yeasts, phototropic bacteria, and lactic acid bacteria then act to transform kitchen scraps and other matter to create rich, dark, fertile earth for gardening or farming.

Boundary layer is a term derived from physics and fluid mechanics to describe a "cushion" of air or strata of fluid that buffers a surface from extremes in temperature, moisture, or momentum transfer. Boundary layers of air exist between the leaf surface and the hairs that emanate from it, thereby reducing the heat load and transpiration of moisture from the leaf itself. Architects are now using this concept to design trellises that reduce heat loads on buildings.

Bunyip is a term favored by permaculture practitioners for homemade levels made from transparent tubes filled with water. There are different designs of these "spirit levels" used for specific applications in a landscape. They are helpful in aligning terraces and other embankments along a contour.

Caliche is a naturally cemented layer or irregular deposition of calcium carbonate formed in soils of arid and semi-arid regions. It may include gravels, sands, and nitrate salts cemented into a hard lime-coated obstacle to digging holes, trenches, and drainage ways in desert landscapes. The roots of both trees and shrubs have difficulty penetrating caliche, and become subject to poor infiltration and oxygen starvation when rare torrential rains fail to percolate or drain away.

Carbon sequestration is a strategy for reducing the volume and consequences of excessive carbon dioxide being released into the atmosphere by the combustion of fossil fuels. Carbon sequestration in the soil has always been done by plant roots, particularly those of woody perennials, but it is now being promoted as a strategy worthy of significant financial investment as a means for slowing global warming.

Check dams are constructed in watercourses to reduce the velocity of storm runoff, impede flooding, and trigger the deposition, settlement, and retention of sediments. As a water-harvesting and erosion control feature, a check dam is typically constructed of rocks, sandbags, or piled debris and strategically placed in a natural or human-made drainage at locations where flood bed-loads can be effectively deposited by breaking the force of moving water.

Coppicing refers to a traditional practice of pruning trees or entire hedges and forests in a manner that produces new shoots and suckers as renewable supplies of timber, forage, or fuelwood without substantively depleting the trees' reserves or impeding their future growth. The noun *coppice* is shorthand for a managed woodland where such pruning has been done as a repeated, sustainable practice.

Desertification is the degradation or impoverishment of arable or habitable lands as a result of inappropriate and unsustainable land uses, or from abiotic stresses aggravated by human-aggravated climate change. The lands lose much of their water-holding capacity and the resilience of their soil, as well as their vegetative cover and microbial diversity.

However, the biotic communities of such degraded or desertified lands do not necessarily gain the composition, structure, or functions of healthy desert communities, but descend to the state of an ecological "disclimax" where future vegetation development is arrested.

Drought evasion or **drought dormancy** is the strategy taken by quick-maturing ephemeral wildflowers and short-cycle annual crops that escape water deficits and other stressful conditions by completing their lives before the stresses arrive. Plants may go dormant as seeds or as inactive underground tubers during extended drought, while animals may estivate, hibernate, or persist only as fertilized, viable eggs until favorable conditions return. Drought tolerance (see below) is the capacity of a plant to survive and stabilize its yields or productivity in the face of water deficits. While drought tolerance is a trait much sought after by plant geneticists, it is typically the multigenic, holistic response of a plant to water stress, rather than a feature based on a single gene that can be simply transferred through biotechnological techniques.

Drought tolerance is the physiological capacity of a plant to endure drought, including soil water deficits, for extended periods of time, without death or severe systemic tissue damage. Plants may osmotically adjust their photosynthetic tissues to maintain their capacity for growth, while other more expendable tissues atrophy under water deficits.

Dry-farming refers to a set of time-tried agricultural techniques for soil moisture retention and food cultivation in semi-arid and arid regions where the economic or ecological costs of conventional irrigated agriculture were prohibitive. A century ago, a series of Dry-Farm Congresses promoted the profitable production of drought-adapted crops that did not require regular, evenly timed irrigation, on lands that received annually a rainfall of 20 inches or less. Ironically, when gas-powered pumps for groundwater extraction became cheaply accessible, much of the wisdom accumulated through the congresses was ignored or even vilified, for commercial farmers abandoned water-efficient crops for hybrids that demanded high inputs of irrigation water and fertilizer to produce yields on the same lands.

Ecosystem services or **nature's services** refer to the benefits humankind derives from work or protective structures and chemical functions provided (usually for free) by natural ecosystems. The provision of clean water and flood control by marshlands, and the provision of wild pollinators by healthy habitats rich in nectar resources, are but two examples of ecosystem services that benefit food production. Ecosystem services are often grouped into four broad categories:

provisioning, such as the generation of potable water and edible wild plants; *regulating*, such as the buffering from weather extremes and prevention of human, crop, or livestock diseases; *supporting*, such as the cycling of nutrients, renewal of soil fertility, and insect pollination; and *cultural*, such as the pursuit of spiritual and recreational renewal in natural environments.

Farmscaping is a holistic ecological approach to pest management, microclimate buffering, and the proliferation of beneficial insects and other wildlife in agrarian landscapes. It involves the use of hedgerows, insectary plants, windbreaks, and bio-swales to diversify the structure of a cultivated environment to enhance food production and protect it from abiotic and biotic stresses. It is promoted by the Wild Farm Alliance and many permaculture associations.

Foodprint is shorthand for the portion of a person's carbon footprint that is generated by the production, processing, transport, storage, and preparation of food.

Foodscape is an agrarian landscape of more than one farm or garden that has been designed to provide the ecosystem provisioning services required for sustained food production.

Foodsheds can be either the schematic representation of geographically distinct food supplies that flow into a locality's food supply chain, or the physical network of farms, ranches, orchards, fisheries, and transportation conduits between them and the consumers served by the aggregation of the foods provided from these sources through one or more food hubs.

A **fredge** is a living fence or boundary marker of trees, shrubs, and herbaceous plants that matures to function as a microclimate-inducing hedge, a silt trap, or a barrier to unwanted livestock and wildlife. Its name combines sounds from the words *fence* and *hedge*.

A **gabion** is a structurally durable wirework or wickerwork container or webbed frame that is filled with cobblestones, broken cement, caliche fragments, or building rubble. It is used singly or in concert with other containers to construct check dams, silt traps, or retaining walls, or to fortify embankments.

Heat tolerance or **thermo-tolerance** refers to the capacity of a plant, animal, or microbe to survive and reproductively function while withstanding periods of excessive heat in the air or on the ground surface. Each food crop has a range of thermo-tolerance in which there is a more bracketed range of optimal conditions for its growth and reproduction.

Honeybee colony collapse disorder is a pathological response to multiple stresses affecting a small but significant percent of honeybee colonies in a dramatic manner: There is a sudden disappearance of worker bees from within and around the vicinity of the hive, which leaves only the queen and newly hatched bees behind. This condition keeps the colony from functioning in pollinator services and in honey production. While many factors have long contributed to the slow but steady decline of honeybees, especially in conventional agricultural fields, the phrase *colony collapse disorder* alerted the public to an abrupt and initially unexplainable decline in honeybee colonies across America that began in late 2006. Since then, European beekeepers have witnessed similarly sudden die-offs of honeybees in 10 countries. However, contrary to the press reports, neonicitinoid insecticides, cell phones, and gmos are not the most important or sole causes of the declines.

Hydro-zones are cultivated areas where plants are grouped for optimal growth according to their similar water needs. These areas are divided into four zones, ranging from high to very low; the lower zone requires less than one-sixth of the irrigation water per year that the highest zone demands, suggesting that considerable conservation of water and monetary savings can accrue from such strategic clustering.

Insectary plants are those that have accentuated features to attract insects with their pollen, nectar, fragrances, as well as their micro-habitats for nesting, roosting, or larval host sheltering. Typically, insectary plants are intentionally introduced into a garden or orchard to increase the specific pollen and nectar resources required by predators, competitors, or parasites of insects considered harmful or unwanted in crop plantings. While many have proven effective in the natural control of pests, these insects may also facilitate the pollination of food crops.

Intercrops are plantings of two or more different crop species within the same row or in alternating rows or other spatial patterns. Their close proximity often generates additive effects such as overyielding.

Micro-irrigation refers to the slow but precise application of water onto the surface of the soil at the base of plants through tiny emitters. These emitters deliver discrete or continuous drips or micro-sprays at frequent intervals for long periods of time, but at relatively low volumes. This moisture is more readily absorbed into the root zone of plants. The best-known (but perhaps only moderately efficient) form of micro-irrigation is drip or trickle irrigation, but other technologies are rapidly emerging.

Mycorrhizae is a term that not only refer to beneficial fungal growths attaching to roots in soil; more properly, it identifies the symbiotic associations essential for one or more partners, among a soil-dwelling (or endolithic) fungus and the roots of one or more species of associated plants. The fungi foster nutrient transfer from the soil to the plants in exchange for carbon fixed by the plants' photosynthetic organs.

Nature's services. See *ecosystem services*.

Nichols bench terraces are variations on broad-based ridge terraces used in arid and semi-arid lands to retain soil and moisture on slopes. The Nichols or channel variation is a graded broad-based terrace for which the soil fill is derived from above the cut line.

Nurse plant guilds are associations focused around a tree or woody shrub that provides shade or other forms of protection from abiotic and biotic stresses. It allows understory plants of its own or other species to germinate and survive beneath its protective canopy, although guild members may eventually outcompete or kill the nurse. The guild itself is any frequently co-occurring and structured group of species beneath the nurse. Each species may provide a unique set of functions that work in conjunction, or harmony, with those offered by the other plants present. They may be linked in subterranean environments through mycorrhizal associations as well. Guilds may not only include groups of plants, but insects, reptiles, birds, and mammals that coexist in the same micro-environment.

Olla (pronounced *OY-yah*) **irrigation** utilizes unglazed clay or terra-cotta pots with a bottle or tapered shape as reservoirs for irrigation water. The pots are buried in the ground with the top exposed above the soil surface and capped. The filled pots can provide subsurface irrigation of plant roots clustered around their moist surface, which seeps moisture out to meet the transpiration demand of the crops. This irrigation technology is truly ancient, with documents from western China suggested centuries of continued use in crop production.

Overyielding is the concept that two food crops intercropped in the same area can collectively produce more yeild than either would do if solely planted in the same space.

Perennial polyculture is the planting of several species of perennial food plants together to grow without additional tillage in the same field or orchard. This ancient strategy has been used for producing both herbaceous and woody perennial food crops on nearly all continents. However, in 1978 Wes Jackson proposed the development of a polyculture of strictly herbaceous perennials—grasses, legumes, and oilseed-bearing sunflowers—that biomimicked but also essentialized the perceived structure of tallgrass prairies. Because his goal was to replace the costly annual production of cereal grains and soybeans he saw around him on the Great Plains, his Land Institute developed simple species mixes that did not need to be resown every year after intensive soil cultivation. His objectives were to leave semi-arid prairie soil structures more intact, to prevent further soil erosion, and to foster symbiotic relationships between the soil and the roots of these perennials. However, attention has become so exclusively focused on breeding perennial cereals and oilseeds that the larger value of mixing woody or herbaceous perennials—including those with edible fruits, nuts, shoots, and leaves—has remained relatively neglected.

Permaculture is an ecological and social framework for consciously designing and sustaining agriculturally productive landscapes that have much of the diversity, stability, and resilience of natural ecosystems. It teaches its practitioners how to apply ecological concepts for the harmonious integration of people's activities in larger cultural landscapes while providing food, fiber, energy, shelter, and other material and nonmaterial needs in a sustainable manner. Initially shaped by Australian visionaries Bill Mollison and David Holmgren, its practices have flourished after being refined and extended by permaculture activists around the world.

Phenological shifts are changes in the timing of growth and reproduction events in the life of an organism, which then alter its relationships with its own as well as other species. The ecological consequences of a phenological shift for a particular individual or population of a food plant will depend upon whether shifts also occur among the other species upon which it relies for pollination or seed dispersal. If food plants and their pollinators respond differently to climate change, then one or both of them may fail at reproduction and suffer population declines.

Pollinator services are a category of ecosystem services that enable sufficient visitation and fertilization of flowers of food and forage crops by insect and other pollen vectors. This service can be provided either by wild native pollinators as one of nature's "free services" or by managed pollinators such as honeybees as a contracted commercial service. There is an ever-growing range of stewardship strategies to enhance pollinator populations on farms to maximize, stabilize, or secure crop quality and food yields. For crop plants that require cross-pollination to produce fruits and seeds, it is essential to sustain viable populations of the 25,000 species of bees documented to live on this earth, not just the domestic honeybee. Fortunately, pollinator habitat enhancement in fredges, vedges, and herb gardens benefits both wild and domesticated bees.

Polyculture is the mixed cropping of species growthforms (eg. trees or herbs) and varieties together; sharing the same vertical and/or horizontal space.

Quadri-modal. See *bimodal*.

Rain gardens are runoff-collecting basins strategically located to capture and retain moisture for the growth of (usually native) medicinal, ornamental, and food plants that utilize otherwise wasted stormwater. Rain gardens are typically constructed where natural depressions can be enhanced to store rainwater running off impervious surfaces such as paved streets, roofs, driveways, sidewalks, and parking lots. One objective of rain garden construction in urban areas is to improve the water quality in nearby bodies of water. Oddly, they have been underutilized for urban food production, especially for fruit-bearing woody perennials.

A **refractometer** can be used to measure the Brix value or soluble sugars (and relative nutrient content) of beverages, fresh vegetables, and fruits. While this unit of measurement was named in 1870 for Adolf F. Brix, the 19th-century scientist who developed this calibration, its significance in assessing the nutrient density of foods as an indicator of soil fertility and health value has only recently been fully appreciated. Technically, the Brix measurement procedure assesses the quantity of the dry matter dissolved in a liquid, which changes the level of light refracted through the liquid being sampled. The dissolution of sucrose and other sugars concentrated in the liquid is a rough indicator of the density of mineral- and vitamin-rich sugars, which offer the most attractive flavor components to most foods and beverages. Brix values are traditionally used in the quality assessment of wines, hard ciders, agave nectars, fruit juices, and honey. The values vary not only with environmental stress and soil fertility levels, but also with preparation and processing techniques used for value-added products.

Retaque fences are constructed of two upright posts at each end of a fence segment interval, with horizontally laid tree logs or palm trunks stacked upon one another to produce a livestock-proof barrier. These are typically comprised of mesquite trunks in the US Southwest and northern Mexico, where they are either used for livestock corrals or as trellises for vedges.

Reverse-sloped terraces are bench terraces constructed in the uplands of more flood-prone regions that are sloped down about 100 mm (about 4 inches) from the highest level of their outer rim or lip. In more arid regions, terraces are often outwardly sloped, which would cause erosion or structural collapse in humid regions where rains are more frequent and sometimes more intense. These terraces are usually planted for the production of dry-farmed or rain-fed crops that require supplemental moisture only during the dry season. They are sloped to allow for drainage under extreme events.

Runoff coefficients are measures of relative amount of precipitation that manifests itself as runoff from the soil, a road, or a roof surface rather than the volume directed into infiltration, export, and evapotranspiration losses. It is framed as the fraction of total rainfall (of a single event or a year) that is dispersed as runoff, once the volume of water stored in a basin or depression, infiltrated into soils, or lost to the atmosphere are accounted for. The runoff coefficient from a particular storm event is calculated as runoff divided by the corresponding rainfall, with both expressed as volumes spread in depths over an acre surface.

Silt traps are sediment deposition basins located behind or below a brush weir, gabion, or fredge that capture soil and flood-washed organic detritus eroded away or washed off from upstream areas during rains. Their nutrient-rich materials can be excavated from the basin and transferred to areas needing soil fertility renewal.

Structural diversity refers to the "architectural" heterogeneity of plants in a forest community, desert oasis plant guild, or orchard. This variety in plant morphology produces many horizontal or vertical physical elements that collectively create a more complex and buffered microclimate for its inhabitants. Trees, shrubs, and vines of different height classes and foliage densities may synergistically benefit one another by providing support structures and protective mechanisms such that the whole of the plant guild is greater than the sum of its parts.

Thermo-tolerance. See *heat tolerance*.

Understory layers of vegetation refers to the plants growing in protected micro-environments beneath the main canopy of a forest or orchard. There may occur multiple strata of plants with different growth forms, from ground-covering vines to upright shrubs that prefer the shade beneath a canopy to the open sun.

The **urban heat island effect** causes heat stress in sprawling metropolitan areas, which routinely suffer warmer nighttime temperatures than their surrounding rural areas. This meteorological and ecological phenomenon was first recognized and investigated by Luke Howard in the 1810s, but largely ignored by most environmental scientists until global warming had dramatically advanced. Ironically, disruptions of temperature, wind, and moisture gradients in urban heat islands do not exactly mirror global weather trends, but appear accentuated or distorted. The primary trigger of the urban heat island effect is extensive urban development, which relies on asphalt, concrete, and metal surfaces that effectively absorb and retain heat. The waste heat generated and expelled by air conditioners, swamp coolers, and refrigerators also contributes to shifting the phenologies and spatial patterns of urban-dwelling organisms.

A **vedge** is a vegetated boundary of a field, garden, or orchard where vines and scandent (sprawling) woody perennial plants overwhelm the trellises, fences, or berms upon which they climb.

Wick irrigation is a highly efficient micro-irrigation technology first pioneered in India, which uses lantern wicks running from water containers to moisten the root zones of clustered plants as they transpire and demand more water. This micro-irrigation technique is far more water conserving and efficient in targeted delivery than drip irrigation, and is suited to cultivated patches in otherwise forbidding or arid environments.

Zones (in permaculture) refers to areas in the foodscape in which certain food-producing organisms are clustered or protected according to the commonalities of their relative needs for management. In inner zones (with lower numbers) closest to the kitchen, the intensity of irrigation management, pruning, feeding, or other stewardship practices requiring frequent intervention is higher. In outer zones, food and beverage resources are minimally maintained, loosely protected, and seldom irrigated or fertilized. The schematic offered below is derived from a number of certified permaculture instructors:

- **Zone 0:** The center or focal point of daily activities such as food preparation—for example, the hearth, a window box of herbs, the pantry, or the canning kitchen.
- **Zone 1:** The most closely guarded and intensively managed cultivated spaces close to the house, including dooryard gardens, cold frames, hoop houses, chicken coops, compost heaps, and controlled environments for fermentation, aquaponics, or mushroom culture.

- **Zone 2:** The mixed orchard-garden of multilayered guilds where fruits, fish, eggs, nuts, greens, and roots are harvested daily or weekly, and food species are nested within intensively cultivated and irrigated beds, on trellises, in ponds, and in poultry runs.
- **Zone 3:** The rain garden, where water harvesting provides moisture for multi-lined mixtures of intercropped cereals and legumes, alley-cropping of fruit and oilseed crops, perennial polycultures, and small pastures for free-ranging small livestock and poultry requiring sheltered sleeping areas.
- **Zone 4:** The extensive and minimally irrigated terraces and hedges of semi-managed and semi-wild food, fiber, and medicinal plants, used for attracting pollinators and providing buffered microclimates to inner zones.
- **Zone 5:** The unmanaged and often sacred wildland spaces from peaks to artesian springs, which provide not only nature's services but also wild inspiration for biomimicry experiments and contemplative renewal—the wellsprings of our spiritual and creative energies, regardless of their material or monetary value.

Notes

Introduction: Wasteland or Food-Producing Oasis

1. William C. Foster. 2012. *Climate and Culture Change in North America AD 900–1600.* University of Texas Press, Austin.
2. Gary Nabhan. 2010. "Profile: Aziz Bousfiha, Desert Mystic or Global Pragmatic?" *Peace Works: Writers in E-Motion.* University of Iowa International Writing Program, Iowa City. www.uipeacework .wordpress.com/2010/05/05.
3. P. C. D. Milly, Julio Betancourt, et al. 2008. "Stationarity Is Dead: Wither Water Management?" *Science* 319: 573–74.
4. James. G. Workman. 2009. *Heart of Dryness: How the Last Bushmen Can Help Us Endure the Coming Age of Permanent Drought.* Walker Publishing, New York.
5. William DeBuys. 2011. *A Great Aridness: Climate Change and the Future of the American Southwest.* Oxford University Press, New York.
6. Jim Corbett. 2005. *Sanctuary for All Life: The Cowbalah of Jim Corbett.* Howling Dog Press, Berthoud, CO.

1. Getting a Grip on Climate Change

1. Anonymous. 2012. "Drought Forces Reductions in US Crop Forecasts." Press release, United Nations World Food Programme, Rome. www.wfp.org. August 10.
2. Blair Fannin. 2012. Updated 2011. "Texas Agricultural Drought Losses Total $7.62 Billion. *Agri-Life Today.* www.today/agrilife /2012/03/21/updated/2011/Texas/agricultural/drought/losses. March 21.
3. Elizabeth Grossman. 2012. "Texas Heat Waves Caused by Global Warming, Says NASA's Hansen." www.insideclimatenews.org. January 21.
4. Peter Backlund, Anthony Janetos, and David Schmiel, editors. 2008. *The Effects of Climate Change on Agriculture, Land Resources, Water Resources, and Biodiversity in the United States.* NOAA Report to US Congress, Washington, DC.

5. Julio L. Betancourt. 2010. *Coping with Non-Stationarity in Water and Ecosystem Management.* US Geological Survey, Tucson, AZ.

6. Benedicta Ward, editor and translator. 1984. *The Sayings of the Desert Fathers: The Alphabetical Collection.* Cistercian Publications, Kalamazoo, MI.

7. John Wortley. 2012. *The Book of Elders: Sayings of the Desert Fathers. The Systemic Collection.* Liturgical Press/Cistercian Publications, Collegeville, MN.

8. Anna Lappé. 2010. *Diet for a Hot Planet: The Climate Crisis at the End of Your Fork and What You Can Do About It.* Bloomsbury Books, New York.

2. *Seeking Inspiration and Solutions from the Time-Tried Strategies Found in the World's Deserts*

1. Thomas E. Sheridan and Gary Paul Nabhan. 1978. "Living with a River: Traditional Farmers of the Rio San Miguel." *Journal of Arizona History* 19(1): 1–16; see also Thomas E. Sheridan. 1988. *Where the Dove Calls: The Political Ecology of a Peasant Corporate Community in Northwestern Mexico.* University of Arizona Press, Tucson.

2. Gary Paul Nabhan and Thomas E. Sheridan. 1977. "Living Fence-rows of the Rio San Miguel, Sonora, Mexico: Traditional Technology for Floodplain Management." *Human Ecology* 5(2): 249–76.

3. Richard T. T. Forman and Jacques Baudry. 1984. "Hedgerows and Hedgerow Networks in Landscape Ecology." *Environmental Management* 8(6): 495–510.

4. Iriana Zuria and J. Edward Gates. 2006. "Vegetated Field Margins in Mexico: Their History, Structure and Function, and Management." *Human Ecology* 34(1): 53–77.

5. Alice Bowe. 2011. *High-Impact, Low-Carbon Gardening: 1001 Ways to Garden Sustainably.* Timber Press, Portland, OR.

6. Zoe Davies and Andrew S. Pulin. 2007. "Are Hedgerows Effective Corridors Between Fragments of Woodland Habitat? An Evidence-Based Approach." *Landscape Ecology* 22(3): 333–51.

7. Iriana Zuria and J. Edward Gates. 2006. "Vegetated Field Margins in Mexico: Their History, Structure and Function, and Management." *Human Ecology* 34(1): 53–77.

8. Gary Paul Nabhan and Thomas E. Sheridan. 1977. "Living Fence-rows of the Rio San Miguel, Sonora, Mexico: Traditional Technology for Floodplain Management." *Human Ecology* 5(2): 249–76.

9. Gary Paul Nabhan and Thomas E. Sheridan. 1977. "Living Fence-rows of the Rio San Miguel, Sonora, Mexico: Traditional Technology for Floodplain Management." *Human Ecology* 5(2): 249–76.

10. Janine Benyus. 1997. *Biomimicry: Innovation Inspired by Nature.* William Morrow and Company, New York.

11. Janine Benyus. 1997. *Biomimicry: Innovation Inspired by Nature.* William Morrow and Company, New York.

12. Janine Benyus. 1997. *Biomimicry: Innovation Inspired by Nature.* William Morrow and Company, New York.

13. Tony Burgess. 1985. "Agave Adaptation to Aridity." Symposium on the Agave, Donald Pinkava, editor. *Desert Plants* 7(2): 39–50.

14. Richard C. Pratt and Gary Paul Nabhan. 1988. "Evolution and Diversity of *Phaseolus acutifolius* Genetic Resources." In Paul Gepts, editor. *Genetic Resources of* Phaseolus *Beans.* Kluwer Publishing, the Netherlands.

15. Gary Paul Nabhan. 1979. "Tepary Beans: The Effects of Domestication on Adaptations to Arid Environments." *Arid Lands Newsletter* 10: 11–16.

16. Hide Omae, Ashok Kumar, and Mariko Shomo. 2012. "Adaptation to High Temperatures and Water Deficit in the Common Bean (*Phaseolus vulgaris* L.) During the Reproductive Period." [Indian] *Journal of Botany* 80414: 1–6.

17. Janine Benyus. 1997. *Biomimicry: Innovation Inspired by Nature.* William Morrow and Company, New York.

18. Wes Jackson, Wendell Berry, and Bruce Coleman, editors. 1985. *Meeting the Expectations of the Land: Sustainable Agriculture.* Counterpoint Press, Berkeley, CA.

19. Gary Paul Nabhan. 1985. "Desert Polycultures." In Wes Jackson, Wendell Berry, and Bruce Coleman, editors. *Meeting the Expectations of the Land: Sustainable Agriculture.* Counterpoint Press, Berkeley, CA.

20. Gary Paul Nabhan. 2009. "Chapter 6: Date Palm Oases and Desert Crops: The Maghreb." In *Where Our Food Comes From: Retracing Nikolay Vavilov's Quest to End Famine.* Island Press, Washington, DC, pp. 84–92.

21. Gary Paul Nabhan. 2007. "Agrobiodiversity Change in a Saharan Desert Oasis, 1919–2006: Historic Shifts in Tasiwit (Berber) and Bedouin Crop Inventories of Siwa, Egypt." *Economic Botany* 61(1): 31–43.

22. Gary Paul Nabhan. 2007. "Agrobiodiversity Change in a Saharan Desert Oasis, 1919–2006: Historic Shifts in Tasiwit (Berber) and Bedouin Crop Inventories of Siwa, Egypt." *Economic Botany* 61(1): 31–43.

23. Gary Paul Nabhan, et al. "Chapter 18: Agrobiodiversity Shifts on Three Continents Since Vavilov and Harlan: Assessing Causes, Processes and Implications for Food Security." In Paul Gepts, et al., editors. *Biodiversity in Agriculture: Domestication, Evolution and Sustainability.* Cambridge University Press, New York, pp. 407–24.

24. Gary Paul Nabhan, J. Garcia, R. Routson, K. Routson, and M. Cariño-Olvera. 2010. "Desert Oases as Genetic Refugia of Heritage Crops: Persistence of Forgotten Fruits in the Mission Orchards of Baja California, Mexico." *International Journal of Biodiversity and Conservation* 2(4): 56–69.

25. Rafael Routson. 2012. *Conservation of Agro-Biodiversity in Baja California Oases.* University of Arizona PhD dissertation, Tucson.

26. Wes Jackson. 1985. *New Roots for Agriculture.* University of Nebraska, Lincoln.

27. David Tilman and John Downing. 1994. "Biodiversity and Stability in Grasslands." *Nature* 367: 363.

3. Will Harvest Rain and Organic Matter for Food

1. Sandra L. Postel. 2011. "Getting More Crop per Drop." In Brian Halweil and Danielle Nierenberg, editors. *2011 State of the World: Innovations That Nourish the Planet.* Worldwatch Institute/W. W. Norton, New York, pp. 39–48.

2. Nancy Laney. 1998. *Desert Water.* Arizona-Sonora Desert Museum Press, Tucson.

3. David Jenkins, 2009. "'When the Well's Dry': Water and the Promise of Sustainability in the American Southwest." *Environment and History* 15(4): 441–62.

4. Anonymous. 2007. "Embedded Water in Food Production." *Institute of Groceries Distribution.* www.idg.com.

5. Sandra L. Postel. 1999. *Pillar of Sand: Can the Irrigation Miracle Last?* Worldwatch Institute/W. W. Norton, New York.

6. Charles F. Hutchinson and Stefanie M. Herrmann. 2008. *The Future of Arid Lands—Revisited.* UNESCO Publishing/Springer-Verlag, Paris.

7. Sandra L. Postel. 2011. "Getting More Crop per Drop." In Brian Halweil and Danielle Nierenberg, editors. *2011 State of the World: Innovations That Nourish the Planet.* Worldwatch Institute/W. W. Norton, New York, pp. 39–48.

8. Marc Reisner. 1986. *Cadillac Desert: The American West and Its Disappearing Water.* Viking Books, New York.

9. John L. Sabo, Tushar Sinha, Laura C. Bowling, Gerrit H. W. Schoups, Wesley W. Wallender, Michael E. Campana, Keith A. Cherkauer, Pam L. Fuller, William L. Graf, Jan W. Hopmans, John S. Kominoski, Carissa Taylor, Stanley W. Trimble, Robert H. Webb, and Ellen E. Wohl. 2010. "Reclaiming Freshwater Sustainability in the Cadillac Desert." *Proceedings of the National Academy of Sciences* 107(50): 21263–69.

10. Andrew Warren. 2006. "Chapter 5: Challenges and Opportunities: Change, Development and Conservation." In Exequiel Escurra, editor. *Global Deserts Outlook.* United Nations Environmental Programme, Nairobi, Kenya, pp. 89–110.

11. Alyson Kenward. 2010. "Revisiting the Cadillac Desert." *On Earth* magazine. www.onearth.org/article/cadillac-desert-revisited. December 20.

12. Jared Diamond. 2005. *Collapse: How Societies Choose to Fail or Succeed.* Penguin Books, New York.

13. Stefanie M. Herrmann and Charles F. Hutchinson. 2006. "Chapter 6: Desert Outlook and Options for Action." In Exequiel Escurra, editor. *Global Deserts Outlook.* United Nations Environmental Programme, Nairobi, Kenya, pp. 112–37.

14. S. Mark Howden, et al. 2007. "Adapting Agriculture to Climate Change." *Proceedings of the National Academy of Sciences* 104(50): 19691–96.

15. Gary Paul Nabhan. 1982. *The Desert Smells Like Rain: A Naturalist in O'odham Country.* University of Arizona Press, Tucson.

16. Maimbo Malesu. 2011. "Rainwater Harvesting." In Brian Halweil and Danielle Nierenberg, editors. *2011 State of the World: Innovations That Nourish the Planet.* Worldwatch Institute/W. W. Norton, New York, pp. 49–50.

17. Gary Paul Nabhan. 1979. "The Ecology of Floodwater Farming in Arid Southwestern North America." *Agroecosystems* 5: 245–59.

18. Gary Paul Nabhan. 1983. "Ak-chin 'Arroyo Mouth' and the Environmental Setting of the Papago Indian Fields in the Sonoran Desert." *Applied Geography* 6(2): 61–75.

19. Nigel Dunnett and Andy Clayden. 2007. *Rain Gardens: Managing Water Sustainably in the Garden and Designed Landscape.* Timber Press, Portland. OR.

20. Roger Bannerman and Ellen Considine. 2003. *Rain Gardens: A How-To Manual for Homeowners.* Wisconsin Department of Natural Resources Publication WT-776, Madison.

21. Estevan Arrellano, editor. 2006. *Ancient Agriculture: Roots and Application of Sustainable Farming by Gabriel Alonso de Herrera.* Ancient City Press, Salt Lake City.

4. *Bringing Water Home to the Root Zone*

1. Gary Pitzer. 2013. *Keeping It Down on the Farm: Agricultural Water Use Efficiency.* Water Education Foundation, Sacramento.

2. Lameck O. Odhiambo, William L. Kranz, and Dean E. Eisenhauer. 2011. "Irrigation Efficiency and Uniformity, and Crop Water Use Efficiency." University of Nebraska Bulletin EC732. University of Nebraska–Lincoln, Lincoln.

3. Gary Nabhan and Naima Taylor. 2004. "Linking Drought and Long-Term Water Scarcity to Food Security in the Four Corners States: A Food Policy Paper." Northern Arizona University Center for Sustainable Environments White Paper, Flagstaff, AZ, pp. 1–36.

4. Grady Gammage Jr., Monica Stigler, David Daugherty, Susan Clark-Johnson, and William Hurt. 2011. *Watering the Sun Corridor: Managing Choices in Arizona's Megapolitan Area.* Arizona State University Morrison Institute for Public Policy, Tempe.

5. Gary Paul Nabhan and Naima Taylor. 2004. Linking Drought and Water Scarcity to Food Security in the Four Corners States. Northern Arizona University Center for Sustainable Environments, White Paper, Flagstaff, AZ, pp. 1–36.

6. Grady Gammage Jr., Monica Stigler, David Daugherty, Susan Clark-Johnson, and William Hurt. 2011. *Watering the Sun Corridor: Managing Choices in Arizona's Megapolitan Area.* Arizona State University Morrison Institute for Public Policy, Tempe.

7. 2012. "Ideas for Colorado River Include a Feeder Pipeline." *New York Times.* December 10, p. A15.

8. Gary Nabhan and Naima Taylor. 2004. "Linking Drought and Long-Term Water Scarcity to Food Security in the Four Corners States: A Food Policy Paper." Northern Arizona University Center for Sustainable Environments White Paper, Flagstaff, AZ, pp. 1–36.

9. Gary Nabhan and Naima Taylor. 2004. "Linking Drought and Long-Term Water Scarcity to Food Security in the Four Corners States: A Food Policy Paper." Northern Arizona University Center for Sustainable Environments White Paper, Flagstaff, AZ, pp. 1–36.

10. David A. Bainbridge. 2002. "Alternative Irrigation Systems for Arid Land Restoration." *Ecological Restoration* 20(1): 23–29.

11. Sam Hitt and Gary Nabhan. 1980. "Pitcher Irrigation for Dry Soil Gardens." *Organic Gardening* 27: 124–27.

12. David A. Bainbridge. 2012. "Buried Clay Pot Irrigation." *Research Notes*. www.sustainabilityleader.org.

13. David A. Bainbridge. 2007. *New Hope for Dry Lands: Guide for Desert and Dry and Restoration*. Island Press, Washington, DC.

14. David A. Bainbridge. 2002. "Alternative Irrigation Systems for Arid Land Restoration." *Ecological Restoration* 20(1): 23–29.

5. *Breaking the Fever*

1. Gary Paul Nabhan and Paul Mirocha. 1985. *Gathering the Desert*. University of Arizona, Tucson.

2. Hide Omae, Ashok Kumar, and Mariko Shomo. 2012. "Adaptation to High Temperatures and Water Deficit in the Common Bean (*Phaseolus vulgaris* L.) During the Reproductive Period." [Indian] *Journal of Botany* 80414: 1–6.

3. G. A. Meehl, T. F. Stocker, W. D. Collins, P. Friedlingstein, A. T. Gaye, J. M. Gregory, A. Kitoh, R. Knutti, J. M. Murphy, A. Noda, S. C. B. Raper, I. G. Watterson, A. J. Weaver, and Z. C. Zhao., S. Solomon, D. Qin, M. Manning, Z. Chen, M. Marquis, K. B. Averyt, M. Tignor, and H. L. Miller, editors. 2007. "Global Climate Projections." In *Climate Change 2007: The Physical Science Basis*. Contribution of Working Group I to the Fourth Assessment Report of the Intergovernmental Panel on Climate Change. Cambridge University Press, Cambridge, UK, and New York.

4. Jim Norwine, John R. Giardino, Gerald R. North, and Juan B. Valdés. 1995. *The Changing Climate of Texas: Predictability and Implications for the Future*. Texas A&M University, College Station.

5. Craig R. Elevitch and John K. Francis. 2006. "*Gliricidia sepium* (Gliricidia)." *Species Profiles for Pacific Island Agroforestry*. www.traditionaltree.org.

6. Meredith L. Driess and Sharon Edgar Greenhill. 2008. *Chocolate, Pathway to the Gods: The Sacred Realm of Chocolate in Mesomerica*. University of Arizona Press, Tucson.

7. Thomas Gage (1775), quoted in Allen M. Young, 1994. *The Chocolate Tree: A Natural History of Cacao*. Smithsonian Institution Press, Washington, DC.

8. Josh J. Tewksbury and J. D. Lloyd. 2001. "Positive Interactions Under Nurse Plants: Spatial Scale, Stress Gradients and Benefactor Size." *Oecologia* 127: 425–34; Joel Flores and Eduardo Jurado. 2003.

"Are Nurse–Protegé Interactions More Common Among Plants from Arid Environments?" *Journal of Vegetation Science* 14: 911–16.

9. Humbert Suzán, Gary Paul Nabhan, and Duncan T. Patton. 1996. "The Importance of *Olneya tesota* as a Nurse Plant in the Sonoran Desert." *Journal of Vegetation Science* 7: 635–44.

10. Jay Withgott. 2000. "Botanical Nursing: From Deserts to Shore-lines, Nurse Effects Are Receiving Renewed Attention." *BioScience* 50(6): 479–84.

11. Gary Paul Nabhan and John L. Carr. 1994. *Ironwood: An Ecological and Cultural Keystone of the Sonoran Desert.* Conservation International/University of Chicago Press, Chicago.

12. Lorena Gómez-Aparicio, Regino Zamora, Jorge Castro, and Jose A. Hodar. 2008. "Facilitation of Tree Saplings by Nurse Plants: Micro-habitat Amelioration or Protection Against Herbivores?" *Journal of Vegetation Science* 19: 161–72.

13. Gary Paul Nabhan. 2004. *Cross-Pollinations: The Marriage of Science and Poetry.* Milkweed Editions, Minneapolis.

14. Gary Paul Nabhan, Mark Slater, and Larry Yarger. 1989. "New Crops for Small Farmers in Marginal Lands? Wild Chiles as a Case Study." In Miguel Altieri and Susanna Hecht, *Agro-Ecology and Small Farm Development.* CRC Press, Boca Raton, FL, pp. 19–26.

6. Increasing the Moisture-Holding Capacity and Microbial Diversity of Food-Producing Soils

1. Another version of Fred's talks on this topic occurs in Fred Kirschenmann. 2010. "Soil Fertility, Food Production and the Faith Community." PlanetShifter.com and openmythsource.com.

2. Stephen M. Ogle, F. J. Breidt, M. D. Eve, and K. Paustian. 2003. "Uncertainty in Estimating Land Use and Management Impacts on Soil Organic Carbon Storage for US Agricultural Lands Between 1982 and 1997." *Global Change Biology* 9:1521–42.

3. Alice Bowe. 2011. *High-Impact, Low-Carbon Gardening: 1001 Ways to Garden Sustainably.* Timber Press, Portland OR.

4. R. Neil Sampson. 1981. *Farmland or Wasteland: A Time to Choose.* Rodale Press, Emmaus, PA.

5. Gene Lodgson. 1975. *The Gardener's Guide to Better Soil.* Rodale Press, Emmaus, PA.

6. M. Charles Gould. 2012. "Compost Increases the Water-Holding Capacity of Droughty Soils." www.msue.amr.msu/news/compost -increases-the-water-holding-capacity-of-droughty-soils. May 12.

7. Ken Meter. 2012. "Finding Food in the Border Counties." *Hungry for Change: Borderlands Food and Water in the Balance.* University of Arizona Southwest Center, Tucson. www.swc.arizona.edu.

8. Mohamed Osman El Sammani and Sayed Mohamed Ahmed Dabloub. 1996. "Making the Most of Local Knowledge: Water Harvesting in the Red Sea Hills of Northern Sudan." In Chris Reij, Ian Scoones, and Camilla Toulmin, editors. *Sustaining the Soil: Indigenous Land and Water Conservation in Africa.* Earthscan, London, pp. 28–34.

9. Suzanne Fish, Paul Fish, and Chris Downum. 1984. "The Marana Community in the Hohokam World." *University of Arizona Anthropological Papers.* Arizona State Museum/University of Arizona Press, Tucson.

10. Grace Gershuny and Joe Smillie. 1999. *The Soul of the Soil: A Soil-Building Guide for Master Gardeners and Farmers.* Chelsea Green Publishing, White River Junction, VT.

11. Grace Gershuny and Joe Smillie. 1999. *Soul of the Soil: A Soil-Building Guide for Master Gardeners and Farmers.* Chelsea Green Publishing, White River Junction, VT.

12. Alice Bowe. 2011. *High-Impact, Low-Carbon Gardening: 1001 Ways to Garden Sustainably.* Timber Press, Portland, OR.

13. Jonathan Bloom. 2010. *American Wasteland: How America Throws Away Nearly Half Its Food.* Da Capo Lifelong Books, Boston.

14. Xiangmin Li Chunxia Xu and Shanmin Su. 2000. "Effect of Deep-Ditch Manuring on Apple Tree Root System in an Arid Farming Orchard." In John M. Laflen, Jungliang Tian, and Chi-Hua Huang, editors. *Soil Erosion and Dryland Farming.* CRC Press, Boca Raton, FL, pp. 73–80.

15. Alice Bowe. 2011. *High-Impact, Low-Carbon Gardening: 1001 Ways to Garden Sustainably.* Timber Press, Portland, OR.

7. *Forming a Fruit and Nut Guild That Can Take the Heat*

1. Anonymous. 2003. "Can Carbon Sequestration Solve Global Warming?" *Science Daily.* www.sciencedaily.com. February 12.

2. Tamsin Kerr. 2008. "Book Review of *The Man Who Planted Trees*." www.helium.com. July 18.

3. Kat West, et al., editors. 2009. *Portland Fruit/Nut Tree Report.* Portland/Multonomah County Food Policy Council, Portland, OR.

4. Alan N. Lakso. 2010. "Estimating the Environmental Footprint of New York Apple Orchards." *New York State Horticultural Society* 18: 26–32.

5. D. A. Kroodsma and C. B. Field. 2006. "Carbon Sequestration in California Agriculture, 1980–2000." *Ecological Applications* 16(5): 1975–85.

6. S. Mark Howden, et al. 2007. "Adapting Agriculture to Climate Change." *Proceedings of the [US] National Academy of Sciences* 104(500): 19691–96.

7. Gary Paul Nabhan. 2009. *Where Our Food Comes From: Retracing Nikolay Vavilov's Quest to End Famine.* Island Press, Washington, DC.

8. Dennis Baldocchi and Simon Wong. 2008. "Accumulated Winter Chill Is Decreasing in the Fruit Growing Regions of California." *Climatic Change* 87(0): 153–66.

9. Eike Luedeling, M. Zhang, and E. H. Girvetz. 2009. "Climatic Changes Lead to Declining Winter Chill for Fruit and Nut Trees in California During 1950–2099." *PLoS ONE* 4(7): e6166. doi:10.1371/journal.pone.0006166.

10. G. P. Nabhan, et al. 2012. "Agrobiodiversity Shifts on Three Continents Since Vavilov and Harlan: Assessing Causes, Processes, and Implications for Food Security." In Paul Gepts, et al., editors. *Biodiversity in Agriculture: Domestication, Evolution and Sustainability.* Cambridge University Press, New York, pp. 407–24.

11. Alice Bowe. 2011. *High-Impact, Low-Carbon Gardening: 1001 Ways to Garden Sustainably.* Timber Press, Portland, OR.

12. Michael Phillips. 2011. *The Holistic Orchard: Trees, Fruits and Berries the Biological Way.* Chelsea Green Publishing, White River Junction, VT.

8. When Terraces Are Edged with Succulents and Herbaceous Perennials

1. Jens Gebauer, Eike Luedeling, Karl Hammer, and Andreas Buerkhart. 2009. "Agro-Horticultural Diversity in the Mountain Oases of Northern Oman." *Acta Horticulturae* 817: 325–32.

2. Edward S. Hyams. 1952. *Soil and Civilization: The Past in the Present.* Thames and Hudson, London.

3. Gary Paul Nabhan. 2008. *Arab/American: Landscape, Culture and Cuisine in Two Great Deserts.* University of Arizona Press, Tucson.

4. Jens Gebauer, Eike Luedeling, Karl Hammer, and Andreas Buerkhart. 2009. "Agro-Horticultural Diversity in the Mountain Oases of Northern Oman." *Acta Horticulturae* 817: 325–32.

5. Eike Luedeling, Jens Gebauer, and Andreas Buerkhart. 2009. "Climate Change Effects on Winter Chill for Tree Crops with Chill Requirements on the Arabian Peninsula." *Climate Change* 96: 219–37.

6. Francisco Arriaga, Greg Andrews, Nick Schneider, and Matt Ruark. 2012. "Soil Erosion After Silage Harvest." *New Horizons in Soil Science* 12(3): 1–2.

7. R. Neil Sampson. 1981. *Farmland or Wasteland: A Time to Choose.* Rodale Press, Emmaus, PA.

8. Rattan Lal, Terry M. Sobecki, Thomas Iivari, and John M. Kimble. 2004. *Soil Degradation in the United States: Extent, Severity, and Trends.* Lewis Publishers/CRC Press, Boca Raton, FL.

9. Rattan Lal, Terry M. Sobecki, Thomas Iivari, and John M. Kimble. 2004. *Soil Degradation in the United States: Extent, Severity, and Trends.* Lewis Publishers/CRC Press, Boca Raton, FL.

10. R. A. Donkin. 1979. *Agricultural Terracing in the Aboriginal New World.* Viking Fund Publications in Anthropology/University of Arizona Press, Tucson.

11. Luuk Dorren and Freddy Rey. 2012. "A Review of the Effect of Terracing on Erosion." *Scape: Soil Erosion and Protection for Europe* 97. www.eusoils.jrc.eu.

12. Luuk Dorren and Freddy Rey. 2012. "A Review of the Effect of Terracing on Erosion." *Scape: Soil Erosion and Protection for Europe* 97. www.eusoils.jrc.eu.

13. Patricia M. Spoerl and George J. Gumerman, editors. 1984. *Prehistoric Cultural Development in Central Arizona: Archaeology of the Upper New River Region.* Southern Illinois University Press, Carbondale.

14. Melissa Kruse-Peeples, Hoski Schaafsma, Katherine A. Spielmann, and John Briggs. 2010. "Landscape Legacies of Prehistoric Agricultural Land Use in the Perry Mesa Region, Central Arizona." In Rebecca Dean, editor. *The Archaeology of Anthropogenic Environments.* Center for Archaeological Investigations, Occasional Paper No. 37, Southern Illinois University, Carbondale, pp. 122–41.

15. Gary Paul Nabhan. 1994. "Finding the Hidden Garden." In *Desert Legends: Re-Storying the Sonoran Borderlands.* Henry Holt, New York, pp. 161–82.

16. Katherine C. Parker, James L. Hamrick, Wendy Hodgson, Dorset W. Trapnell, Albert Parker, and Robert Kuzoff. 2007. "Genetic Consequences of Pre-Columbian Cultivation for *Agave murpheyi* and *A. delamateri* (Agavaceae)." *American Journal of Botany* 94(9): 1479–90.

17. Ted C. Sheng. 2000. "Terrace System Design and Application Using Computers." In John M. Laflen, Junliang Tian, and Chi-Hua Huang, editors. *Soil Erosion and Dryland Farming.* CRC Press, Boca Raton, FL, pp. 381–90.

18. Ted C. Sheng. 2002. "Bench Terrace Construction Made Simple." *Twelfth ISCO Conference Proceedings.* International Soil Conservation Organization with Tsinghua Universiuty Press, Beijing.

19. Charles McRaven. 2007. *Stone Primer.* Storey Publishing, North Adams, MA.

9. *Getting Out of the Drought*

1. Gary Paul Nabhan. 2009. *Where Our Food Comes From: Retracing Nikolay Vavilov's Quest to End Famine.* Island Press/Shearwater Books, Washington, DC.

2. Tom Philpott. 2012. "Climate Change Is Already Shrinking Crop Yields." *Mother Jones.* www.motherjones.com. July 4.

3. Gary Nabhan and Carol Thompson. 2010. "Seed Money." *The Economist.* www.economist.com. December 16.

4. Tom Philpott. 2012. "Superinsects Are Thriving in This Summer's Drought." *Mother Jones.* www.motherjones.com. August 8.

5. Kevin Bullis. 2012. "Is Climate Change to Blame for the Current US Drought?" *Technology Review.* www.technologyreview.com. July 31.

6. Gary Nabhan and Carol Thompson. 2010. "Seed Money." *The Economist.* www.economist.com. December 16.

7. Janisse Ray. 2012. *The Seed Underground: A Growing Revolution to Save Food.* Chelsea Green Publishing, White River Junction, VT.

8. Brenda Lin. 2011. "Resilience in Agriculture Through Crop Diversification: Adaptive Management for Environmental Change." *BioScience* 61(3): 181–93.

9. Dominique Desclaux, Salvatorre Ceccarelli, John Navazio et al. 2012. "Centralized or decentralized breeding: the potentials of participatory approaches for low-input and organic agriculture. In Edith T. Lammerts van Bueren and James A. Myers, editors. *Organic Crop Breeding.* New York, Wiley-Blackwell, pp. 99–120.

10. Gary Paul Nabhan, editor. 2008. "Gaspé Flint Corn." In *Renewing America's Food Traditions.* Chelsea Green Publishing, White River Junction VT, pp. 116–17.

11. L. Balaguer, F. I. Pugnaire, E. Martínez-Ferri, C. Arma, F. Valladares, and E. Manrique. 2002. "Ecophysiological Significance of Chlorophyll Loss and Reduced Photochemical Efficiency Under Extreme Aridity in *Stipa Tenacissima* L." *Plant and Soil* 240: 343–52.

12. J. W. Chandler and Dorothea Bartelsand. 2007. "Chapter 55: Drought: Avoidance and Adaptation." In Stanley Trimble, editor. *Encyclopedia of Water Sciences.* CRC Press, Boca Raton, FL, pp. 222–24.

13. Alice Bowe. 2011. *High-Impact, Low-Carbon Gardening: 1001 Ways to Garden Sustainably.* Timber Press, Portland, OR.
14. Jane Mt. Pleasant. 2010. "Estimating Productivity of Traditional Iroquois Cropping Systems Using Field Research and Historical Literature." *Journal of Ethnobiology* 30: 52–79.
15. Suzanne Nelson. 1994. *Genotype and Cropping System Effects on Cowpea Growth and Yield.* University of Arizona PhD dissertation, Tucson. 152 pp. See also Suzanne C. Nelson, Gary P. Nabhan, and Robert H. Robichaux. 1991. "Effects of Water, Nitrogen and Competition on Growth, Yield and Yield Components of Field-Grown Tepary Bean." *Experimental Agriculture* 27(2): 211–15.

10. Getting in Sync

1. Jessica R. Forrest and James D. Thomson. 2011. "An Examination of Synchrony Between Insect Emergence and Flowering in Rocky Mountain Meadows." *Ecological Monographs* 81(3): 469–91.
2. Stein Joar Hegland, Anders Nielson, Amparo Lazaro, and Anne-Line Bjernes. 2009. "How Does Climate Warming Affect Plant–Pollinator Interactions?" *Ecology Letters* 12: 1284–1295.
3. Jane Memmott, P. G. Craze, Nick Waser, and Mary Price. 2007. "Global Warming and the Disruption of Plant–Pollinator Interactions." *Ecology Letters* 10: 710–17.
4. Camille Parmesan, Nils Ryrholm, Constantí Stefanescu, Jane K. Hill, Chris D. Thomas, Henri Descimon, Brian Huntley, Lauri Kaila, Jaakko Kullberg, Toomas Tammaru, W. John Tennent, Jeremy A. Thomas, and Martin Warren. 1999. "Poleward Shifts in Geographical Ranges of Butterfly Species Associated with Regional Warming." *Nature* 399: 579–83.
5. David Inouye. 2011. "Where Have All the Flowers Gone? High-Mountain Wildflower Season Reduced, Affecting Pollinators Like Bees, Hummingbirds." *Science Daily.* June 17.
6. Ignasi Bartomeus, John S. Ascher, David Wagner, Bryan N. Danforth, Sheila Coll, Sarah Kornbluth, and Rachael Winfree. 2011. "Climate-Associated Phenological Advances in Bee Pollinators and Bee-Pollinated Plants." *Proceedings of the National Academy of Sciences* 108(21): 20645–49.
7. Sydney A. Cameron, Jeffrey D. Lozier, James P. Strange, Jonathan B. Koch, Nils Cordes, Leellen F. Solter, and Terry L. Griswold. 2011. "Patterns of Widespread Decline in North American Bumble Bees." *Proceedings of the National Academy of Sciences.* www.pnas.org /lookup/suppl/doi:10.1073/pnas.1014743108/-/DCSupplemental.

8. S. G. Potts, J. C. Biesmeijer, Clare Kremen, P. Neumann, O. Schweiger, and W. E. Kunin. 2010. "Global Pollinator Declines: Trends, Impacts and Drivers." *Trends in Ecology and Evolution* 25: 345–53.

9. Stephen Buchmann and Gary Nabhan. 1996. *The Forgotten Pollinators.* Island Press, Washington, DC; and Gary Nabhan. 2007. Introduction to Ross Conrad, *Natural Beekeeping: Organic Approaches to Modern Apiculture.* Chelsea Green Publishing, White River Junction, VT.

10. Eric Mader, Marla Spivak, and Elaine Evans. 2010. *Managing Alternative Pollinators: A Handbook for Beekeepers, Growers and Conservationists.* SARE Handbook 11, NRAES, Ithaca, NY.

11. Harvey Morris. 2012. "Honey Producers Lament a Bad Season for Bees." *New York Times.* September 29.

12. Gary Paul Nabhan. 2010. "Perspectives in Ethnobiology: Ethnophenology and Climate Change." *Journal of Ethnobiology* 30(1): 1–4.

13. Jessica R. K. Forrest and James D. Thomson. 2011. "An Examination of Synchrony Between Insect Emergence and Flowering in Rocky Mountain Meadows." *Ecological Monographs* 81(3): 469–91.

14. Rachael Winfree, Neal M. Williams, Jonathan Dushoff, and Claire Kremen. 2007. "Native Bees Provide Insurance Against Ongoing Honey Bee Losses." *Ecology Letters* 10: 1105–13.

15. James Cane and collaborators. 2005. "Importance of Squash Bees (*Peponapis & Xenoglossa*) as Pollinators of Domestic *Cucurbita* in the Americas: SPAS (Squash Pollinators of the Americas Survey)." Utah State University/USDA/ARS, Logan.

16. Gary Paul Nabhan. 2001. "Nectar Trails of Migratory Pollinators: Restoring Corridors on Private Lands." *Conservation in Practice* 2: 20–26.

17. L. A. Morandin and M. L. Winston. 2006. "Pollinators Provide Economic Incentive to Preserve Natural Land in Agroecosystems." *Agriculture, Ecosystems and the Environment* 116: 289–92.

18. Rachel F. Long and John Anderson. 2010. "Establishing Hedgerows on Farms in California." University of California at Davis Agriculture and Natural Resources Publication 8390, Davis. http:anrcatalog.ucdavis.edu.

19. Mariken Kjohl, Anders Nielsen, and Nils Christian Stenseth. 2011. "Potential Effects of Climate Change on Crop Pollination." In *Pollination Services for Sustainable Agriculture.* UN Food and Agriculture Organisation, Rome, pp. 2–49.

20. Eric Mader, Matthew Shepherd, Mace Vaughan, Scott Hoffman Black, and Gretchen LeBuhn. 2011. *Attracting Native Pollinators:*

Protecting North America's Bees and Butterflies. Xerces Society/
Storey Publishing, North Adams, MA.

Afterword: Creating Your Own Sowing Circle

1. Masanobu Fukuoka. 2011. *Sowing Seeds in the Desert: Natural Farming, Global Restoration and Ultimate Food Security.* Chelsea Green Publishing, White River Junction, VT.
2. Michael Bomford. 2009. "How Much Greenhouse Gas Comes from Food?" www.energyfarms.wordpress.com. April 16. See also Richard H. Heinberg and Michael Bomford. 2009. *The Food and Farming Transition: Toward a Post Carbon Food System.* The Post Carbon Institute, Santa Rosa, CA.
3. Anna Lappé. 2010. *Diet for a Hot Planet: The Climate Crisis at the End of Your Fork and What You Can Do About It.* Bloomsbury Books, New York.
4. Courtney White, et al. 2004. "Finding the Radical Center: A Manifesto." www.quiviracoaltion.org.
5. Rainier Maria Rilke. 1996. "Alles wird weider gross." In *Rilke's Book of Hours: Love Poems to God,* translated by Anita Barrows and Joanna Macy. Riverhead Books/Penguin Group USA, New York.

Index

About the Author

DENNIS MORONEY

Gary Paul Nabhan is an internationally celebrated nature writer, food and farming activist, and proponent of conserving the links between biodiversity and cultural diversity. He has been honored as a pioneer and creative force in the "local food movement" and seed-saving community by *Utne Reader*, *Mother Earth News*, Bioneers, the *New York Times*, and *Time* magazine. As the W.K. Kellogg Endowed Chair in Sustainable Food Systems at the University of Arizona Southwest Center, he works with students, faculty, and nonprofits to build a more just, nutritious, sustainable, and climate-resilient foodshed spanning the US-Mexico border. He was among the earliest researchers to promote the use of native foods in preventing diabetes, especially in his role as a cofounder and researcher with Native Seeds/SEARCH. Gary is also personally engaged as an orchard-keeper, wild-foods forager, and pollinator-habitat restorationist, working from his small farm in Patagonia, Arizona, near the Mexican border. He has helped forge "the radical center" for collaborative conservation among farmers, ranchers, indigenous peoples, and environmentalists in the West. Follow Gary's lectures, workshops, and blogs at www.garynabhan.org and at the pollinator recovery alliance he cofounded, www.makewayformonarchs.org.